Photovoltaic Systems Technology

Scrivener Publishing
100 Cummings Center, Suite 541J
Beverly, MA 01915-6106

Publishers at Scrivener
Martin Scrivener (martin@scrivenerpublishing.com)
Phillip Carmical (pcarmical@scrivenerpublishing.com)

Photovoltaic Systems Technology

Edited by
Mohammed Aslam Husain
Md Waseem Ahmad
Farhad Ilahi Bakhsh
P. Sanjeevikumar
and
Hasmat Malik

Scrivener Publishing

WILEY

This edition first published 2024 by John Wiley & Sons, Inc., 111 River Street, Hoboken, NJ 07030, USA and Scrivener Publishing LLC, 100 Cummings Center, Suite 541J, Beverly, MA 01915, USA

For more information about Scrivener publications please visit www.scrivenerpublishing.com.

Wiley Global Headquarters
111 River Street, Hoboken, NJ 07030, USA

For details of our global editorial offices, customer services, and more information about Wiley products visit us at www.wiley.com.

Library of Congress Cataloging-in-Publication Data

ISBN 9781394166428

Front cover images supplied by Adobe Firefly
Cover design by Russell Richardson

Set in size of 11pt and Minion Pro by Manila Typesetting Company, Makati, Philippines

Printed in the USA

10 9 8 7 6 5 4 3 2 1

Contents

List of Contributors

R.C. Bansal, Department of Electrical Engineering, University of Sharjah, Sharjah, United Arab Emirates and Department of Electrical Engineering, University of Pretoria, South Africa, rcbansal@ieee.org

Seerin Ahmad, Department of Electrical Engineering and Computer Science, Texas A&M University, Kingsville, USA, seerin.ahmad@students.tamuk.edu

Hwa-Dong Liu, Department of Electrical Engineering, National Taiwan University of Science and Technology, Taipei, Taiwan, hdliu@ntnu.edu.tw

Bushra Sabir, Department of Electrical Engineering, National Taiwan University of Science and Technology, Taipei, Taiwan, bushra11sabir@gmail.com

Rashid Ahmed Khan, Department of Electrical Engineering, National Taiwan University of Science and Technology, Taipei, Taiwan, rashidkhan6417@gmail.com

Sarosh Patel, School of Engineering, University of Bridgeport, CT, USA, saroshp@bridgeport.edu

Mohd Fazil, Centre for Transformative Learning, University of Limerick, Limerick, Ireland, mohd.fazil@ul.ie

Saqib Qamar, Department of Physics, Umea University, Linnaeus vag 20, Umea, Sweden, saqib.qamar@umu.se

Farhad Muhsin Mahmood, Department of Physics, College of Science, University of Halabja, Halabja, Iraq, farhad.mahmood@uoh.edu.iq

Jian Ping Li, School of Communication and Information Engineering, University of Electronic Science and Technology of China, Chengdu, China, lj001@uestc.edu.cn

Shubham Tiwari, Department of Sustainable Energy Transition, School of Environment, Resources and Development, Asian Institute of Technology, Thailand, tiwariucestudy@gmail.com

Vaishali Gautam, Department of Electrical Engineering, Jamia Millia Islamia, New Delhi, India, vgautam603@gmail.com

Shahida Khatoon, Department of Electrical Engineering, Jamia Millia Islamia, New Delhi, India, shahidakhatoon2008@gmail.com

Mohd Faisal Jalil, Department of Electrical Engineering, Aligarh Muslim University, Aligarh, India, mfaisaljalil@gmail.com

Mohammad Shahabuddin, Department of Electrical Engineering, Integral University, Lucknow, India, mshahab@iul.ac.in

Mohammed Asim, Department of Electrical Engineering, Integral University, Lucknow, India, masim@iul.ac.in

Adil Sarwar, Department of Electrical Engineering, Aligarh Muslim University, Aligarh, India, adil.sarwar@zhcet.ac.in

Noorul Islam, Department of Electrical Engineering, Aligarh Muslim University, Aligarh, India, noorulislam40506@gmail.com

Dushyant Sharma, Department of Electrical Engineering, Aligarh Muslim University, Aligarh, India, indushyant20@gmail.com

Mohd Shariz Ansari, Department of Electrical & Electronics Engineering, KIET Group of Institutions, Delhi-NCR Ghaziabad, India, mohdshariz@gmail.com

Akhlaque Ahmad Khan, Department of Electrical Engineering, Integral University, Lucknow, India, akhlaque@iul.ac.in

Ahmad Faiz Minai, Department of Electrical Engineering, Integral University, Lucknow, India, fzminai@gmail.com

Ahmad Neyaz Khan, Department of Computer Science and Application, KL University, Vijayawada, India, ahmadnk500@gmail.com

Asif Khan, Department of Computer Application, Integral University, Lucknow, Uttar Pradesh, India, asif05amu@gmail.com

Asad Malik, School of Information Technology, Monash University, Subang Jaya, Selangor, Malaysia, asad.malik@monash.edu

Gulnar Perveen, Defence Research & Development Organisation, Metcalfe House Annexe, Delhi, India, gulegulnar@gmail.com

Priyanka Anand, Dept. of Electronics & Communication Engg., B.P.S. Mahila Vishwavidyalaya, Sonipat, Haryana, India, anand_priyanka10@yahoo.co.in

Amod Kumar, Defence Research & Development Organisation, Metcalfe House Annexe, Delhi, India, amod.drdo@gmail.com

Mohd Khursheed, Department of Electrical Engineering, Integral University, Lucknow, Uttar Pradesh, India, khursheed20@gmail.com

Farhan Ahmad, Department of Computer Application, Integral University, India, farhanlucknowi@gmail.com

Satya Prakash Makhija, Department of Electrical Engineering, Vindhya Institute of Technology and Science, Satna, India, satyapmakhija@gmail.com

Pankaj Kumar Shrivastava, Department of Mechanical Engineering, Government Engineering College, Bhojpur (Bihar), India, psiitd@yahoo.com

Prasanta Kumar Jena, Department of Electrical and Electronics Engineering, BVRIT Hyderabad College of Engineering for Women, Hyderabad, Telangana, India, pkj.ped@gmail.com

Satya Prakash Dubey, ICFAI Universality, Raipur, Chhattisgarh, India, spdubeyfee@gmail.com

Pushpendra Singh, Department of Electrical Engineering, Rajkiya Engineering College, Banda, Uttar Pradesh, India, erpsingh@recbanda.ac.in

Dileep Kumar, Department of Electrical and Electronics Engineering, Motihari College of Engineering, India, dk18182@gmail.com

Md Tabrez, Department of Electrical and Electronics Engineering, Motihari College of Engineering, India, mdtabrez2008@gmail.com

Surya Deo Choudhary, Department of Electrical and Electronics Engineering, Motihari College of Engineering, India, suryadeo.bit@gmail.com

Rupendra Kumar Pachauri, Electrical Cluster, School of Engineering, University of Petroleum and Energy Studies, Dehradun, rupendra-pachauri@gmail.com

Renu Mavi, Department of Chemistry, Swami Vivekanand Subharti University, Meerut, Uttar Pradesh, drrenumavi@gmail.com

Shashikant, Electrical Engineering Department, Babu Banarasi Das University, Lucknow, Uttar Pradesh, shashikant52@gmail.com

Preface

As of 2022, the cumulative installed capacity of solar photovoltaic (PV) systems has surpassed 1,185 GW, representing approximately 5% of the global energy output. Grid-connected systems account for 60%, whereas the rest are distributed applications. By 2030, 80% of the electrical power is expected to flow through power electronics. The processing and reliability of power from PV systems have attracted the attention of researchers. Many research articles and books have been published in the last two decades, covering different issues of PV efficiency, circuits, systems for power processing, and their related control. Books that have been published cover one or more topics but fail to provide a complete picture of the different aspects of PV systems. The aim of this book is to fill this gap by providing a comprehensive review of techniques and practices dedicated to improving the performance of PV systems.

Because the performance of PV systems depends on better estimates of solar irradiance, extraction of maximum power available from PV panels, and optimization of the design of power electronics converters for energy processing, this book intends to cover these aspects. This book comprised of 12 different chapters.

Chapter 1 discusses the historical developments of solar PV systems. A brief introduction to different PV models is presented. Chapter 2 describes the current developments in solar PV systems. A brief introduction to the different PV inverter architectures is presented. Recent developments in microinverters and challenges in DC–DC converters for PV applications are also presented.

Chapter 3 reviews different reconfiguration approaches for the clustering of panels to maximize the performance of the PV array. Both the static and dynamic reconfiguration strategies are discussed, and their comparative performances are presented.

In Chapter 4, various isolated and nonisolated DC–DC converters used in PV systems are discussed. Their comparative performance evaluation is done.

Chapter 4 deals with different maximum power point tracking (MPPT) algorithms used in PV systems. Approaches such as P&O, IC, ANN-based MPPT, FLC-based MPPT, GMPP, PSO, and other metaheuristic-based MPPT algorithms applied to PV systems are discussed in this chapter.

In Chapter 5, a comprehensive study of conventional, meta-heuristic, and AI-based MPPT techniques is presented with respect to their complexity, tracking speed, steady-state fluctuation, etc. The benefits and drawbacks of MPPT methods are also tabulated.

Chapter 6 discusses the different artificial intelligence (AI) techniques and methods used in the modeling and prediction of solar radiation data. To optimize solar radiation data forecasting, pseudocodes of ACO, SVM, and genetic algorithms are provided. Different deep learning, ANN, and expert systems are also discussed.

Chapter 7 proposes an ANN–ANFIS model using meteorological parameters for forecasting solar power. Four models of different meteorological conditions were developed in MATLAB, and it was concluded that hazy models are more accurate than the other produced models.

Chapter 8 proposes an ARIMA model for solar PV forecasting. The forecasting performance of the implemented ML algorithms is compared with that of the Smart Persistence (SP) method and the research shows that the ML models outperform.

A case study for evaluating the techno-economic feasibility of ground and floating PV Systems is presented in Chapter 9. The System Advisory Model was used to simulate a 2,500 MW solar plant for both ground and floating PV systems. Thus, floating PV systems are financially more feasible.

The application of solar PV to water-pumping systems is discussed in Chapter 10. The MATLAB/Simulink model was developed for a BLDC motor that drives a centrifugal pump. Various operating modes and resulting waveforms are also provided.

Chapter 11 introduces a hybrid PV/PEMFC-driven water-pumping system for agricultural applications. A MATLAB/Simulink model is developed to study the steady-state and transient responses of the system. Various aspects of the utilization and integration of hybrid PV/PEMFCs are discussed.

Chapter 12 provides a comprehensive study on the worldwide deployment of solar PV-EV systems and modern solar-assisted electric vehicle charging systems. This chapter discusses the evolution of EV, various types of charging stations, and various innovative energy-storage methods.

This book is a valuable resource for individuals interested in the modeling, design, forecasting, and application of solar photovoltaic (PV) systems using metaheuristic, evolutionary computation, machine learning, and

AI approaches. Researchers and practicing engineers alike will find this book highly adaptable and beneficial for their work. With its wide coverage of subject areas and wealth of information ranging from fundamental to advanced levels, this single volume is a complete guide for anyone interested in this field.

We express our sincere gratitude to all contributors to this book for their invaluable efforts in creating exceptional literature for the research community. Their dedication and expertise have resulted in the production of high-quality content that will undoubtedly benefit and enrich research.

We thank our colleagues and friends for their insightful and helpful discussions during the production of this edited book.

Mohammed Aslam Husain
Dept. of Electrical Engg., REC Ambedkar Nagar, AKTU, India
Md Waseem Ahmad
EEED, NIT Karnataka, Surathkal, Karnataka, India
Farhad Ilahi Bakhsh
EED, NIT Srinagar Hazratbal, India
P. Sanjeevikumar
Aarhus BSS, Aarhus University, Herning, Denmark
Hasmat Malik
Department of Electrical Power Engineering, Faculty of Electrical Engineering, Universiti Teknologi Malaysia (UTM), Malaysia

1

History of Solar PV System and its Recent Development

Vaishali Gautam[1], Shahida Khatoon[1] and Mohd Faisal Jalil[2*]

[1]*Department of Electrical Engineering, Jamia Millia Islamia, New Delhi, India*
[2]*Department of Electrical Engineering, Aligarh Muslim University, Aligarh, Uttar Pradesh, India*

Abstract
Solar PV systems are becoming increasingly important in compensating for the shortage of electrical energy caused by rising demand and decreasing conventional energy sources. The level of carbon dioxide emissions surpassed the 440-ppm limit in 2017, resulting in an uncontrolled climate shift. To address the increasing need for decarbonization, wind energy (WE) and photovoltaic (PV) systems are being installed more frequently in developed and developing countries. PV systems are the most straightforward, reliable, and clean way to generate power from solar radiation. The photovoltaic (PV) effect was first observed by Alexandre Edmond Becquerel in 1839, and the first PV cell with a low efficiency of 6% was developed in 1954, which has now increased to 20%–22%. Examining the history of PV cells can provide valuable insights to guide future developments.

Keywords: Partial shading condition, PV array, reconfiguration, MPPT

1.1 Introduction

Owing to new enthusiastic tendencies that promote environmental preservation and the ongoing rise in energy consumption, renewable power production technologies have emerged as a significant study area. The rate of growth in the world's energy demand in 2018 was 2.3%, which was the highest rate in the last 10 years [1]. Owing to their detrimental ecological

**Corresponding author*: mfaisaljalil@gmail.com

Mohammed Aslam Husain, Md Waseem Ahmad, Farhad Ilahi Bakhsh, P. Sanjeevikumar and Hasmat Malik (eds.) Photovoltaic Systems Technology, (1–26) © 2024 Scrivener Publishing LLC

impact and escalating costs, sources reliant on petroleum derivatives need to be used less frequently [2–4].

The year 2017 marked a turning point in the production of renewable energy. Surprisingly, 178 GW of additional renewable installations represent more than 66% of the world's net power capacity. At 97 GW, solar photovoltaic (PV) installations increased the most, with China accounting for over half of this growth [5, 6]. According to the International Energy Agency (IEA) forecasts, an increase of approximately 1 TW, or a 46% in renewable power output is predicted for 2018 and 2023. Approximately half of this growth is accounted for by photovoltaic (PV) power, which is driven by supportive market and government policies [7]. The growth in the production of renewable energy is shown in Figure 1.1, and the growth in the production of PV electricity is shown in Figure 1.2.

Wind continues to be the 2nd largest donor in the development of renewable capability in this forecast, after hydropower and bioenergy. It is anticipated that wind capacity will increase by 60% (approximately 324 GW), with offshore wind accounting for 10% of this growth. Most of the time, hydropower and bioenergy development prospects are somewhat more optimistic than they were a year ago. This is probably due to developments in China [8, 9].

The IEA considers a different prediction situation, known as the favored case. This often-used example of prediction describes how market and methodological advancements can affect renewable installations. By 2023, renewable energy installations could grow by 25% more than in the preceding scenario, reaching 1.3 TW. According to the prediction of

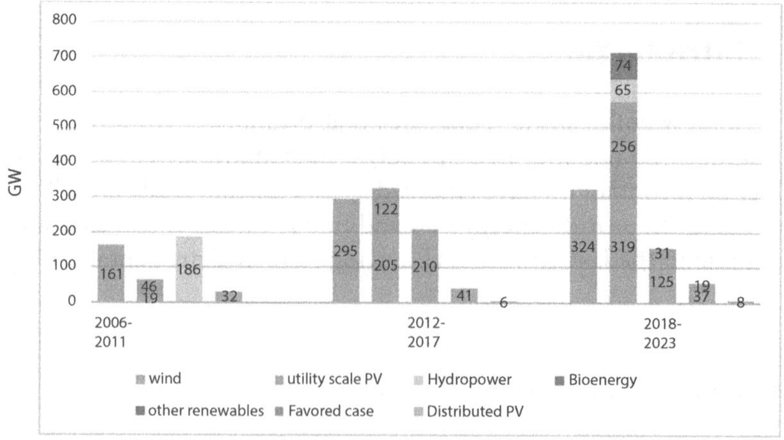

Figure 1.1 Growth in renewable power generation.

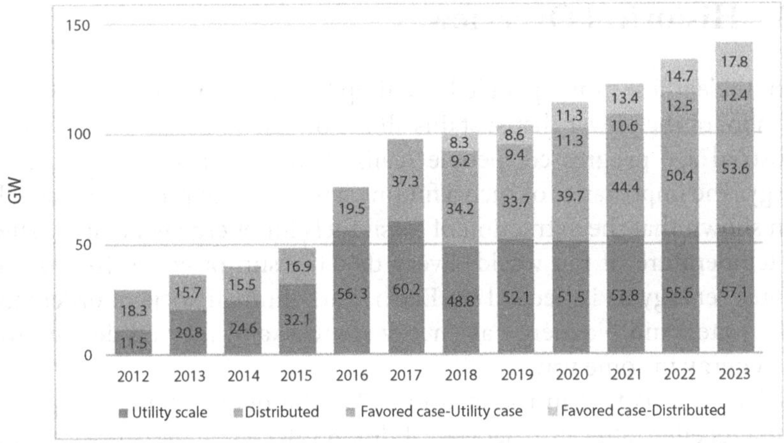

Figure 1.2 Growth in PV power generation.

the favored case, China, India, Europe, and the United States account for approximately 66% of new installations [10, 11].

1.2 Solar Photovoltaic (PV)

In the past two decades, the global rise in photovoltaics (PVs) has been virtually exponential. Photovoltaic (PV) installations are expected to grow the most in the next six years, with 575 GW of new installations coming online. In addition to the rapid growth of distributed generation installations, utility-scale ventures account for 55% of this expansion [12].

In favored case prediction, photovoltaics alone contributed to half of the additional development. The annual increase required to reach 140 GW by 2023 is driven by faster cost decreases, which makes the technology more focused everywhere. Most new developments in the field of photovoltaics consist of residential, commercial, and off-grid applications. This suggests a significant amount of untapped potential in these areas, particularly in countries such as China, India, Europe, and the United States [7].

India is the third fastest growing country in solar power business after China and the United States [7, 13]. In 2010, the National Solar Mission of India established a target of generating 20 GW of solar energy by 2022. This goal was accomplished four years earlier, in January 2018. The Indian government established a new goal for renewable power generation by 2022. Solar power generation of 100 GW and renewable energy output of 175 GW constitute the new goal [14].

1.3 Historical Overview

From the earliest times, people looked up to the sun as a divine being and a symbol of the power that sustains all life on Earth. Later, as learning and industrial age progressed, people realized that the sun was a source of energy. The importance of such a finding has peaked at present, when it has been shown that the extraction of fossil fuels for energy generation affects the temperature of the world. Every day, the sun provides 10,000 times as much energy as is needed on Earth. Early human societal orders used wind, water, and bioenergy as energy sources, and they all derived from solar energy in some way.

Solar PV was first introduced in 1839. The photovoltaic effect, or the Becquerel effect after its discoverer Edmond Becquerel, was first observed by him when he was 19 years old. When he shone light upon an electrode submerged in an acidic liquid, he found that the current flowed through the electrodes [15]. The term "Father of Photovoltaics" was later attributed to him.

No major progress has been made in this field between 1839 and 1877. Adams and Day created a selenium PV cell in 1877 [16]. In 1883, Fritts first proposed using a thin layer of selenium as a solar cell. Although Fritts thin-film selenium solar cells have been thoroughly tested and implemented, the scientific community is still doubtful. A lack of faith in the idea persisted, even though the experiments could be repeated and reproduced. Simply put, the hypothetical operating principle of solar cells was beyond the scope of conventional physics back in those days [17].

Significant progress was made in this key area in 1900. As a result, Quantum Mechanics was developed. At the same time, Max Planck introduced the concept of energy packets, or quanta, with his now-famous equation $E = h.v$ [18]; he also introduced the idea of a photon. Albert Einstein introduced the concept of photon packets or light quanta in an article published in Annalender Physik in 1905. This process was the foundation upon which semiconductor engineering was built as it provided a complete explanation of the PV concept.

Grondahl conducted research and wrote numerous articles on copper cuprous oxide solar cells in 1933 [19]. However, Ohl filed the first silicon solar-cell patent in 1941. Considering its low efficiency (less than 1%), this silicon solar cell is not marketable. However, in terms of PV cells, this is a significant step forward [20].

Chapin, Fuller, and Pearson described a 6% efficient silicon solar cell in a 1954 article in the Journal of Applied Physics [21]. This silicon PV cell was used as this as its basis. Since then, numerous advances have paved the way for new solar cell designs. Currently, a wide variety of solar cells are available in the market, but researchers are always looking to improve their technology and find new ways to harness the sun's rays for use. Advances in structures and materials are crucial for the development of such technologies. In any case, the goal of maximizing efficiency while minimizing expenditure has not changed.

According to the most recent data [22, 23], monocrystalline solar cells have a lab efficiency of 26.7%, whereas their multi-crystalline counterparts only achieve 22.3%. Copper indium gallium selenide (CIGS) and cadmium telluride (CdTe) thin-film solar cells achieved record-high laboratory efficiencies of 22.9% and 21.0%, respectively. Over the past decade, the efficiency of commercial silicon crystalline modules has increased from 12% to 17%. The commercial CdTe-based PV module efficiencies increased from 9% to 16% over the same period. Multijunction solar cells have achieved an efficiency of 46% in the laboratory.

1.4 Grid-Connected PV System

Our ability to use the sun's energy for a wide variety of tasks is made possible by the PV system, which consists of several interconnected components. Typically, a PV system has two primary parts: PV modules and an inverter. Battery storage is also necessary; its exact role varies depending on the system design [24].

Countries with robust solar energy policies tend to prefer grid-connected configurations, as they enable consumers to supply surplus power to the utility company. The PV module and inverter are the two main components of the grid-connected systems. The PV modules are responsible for

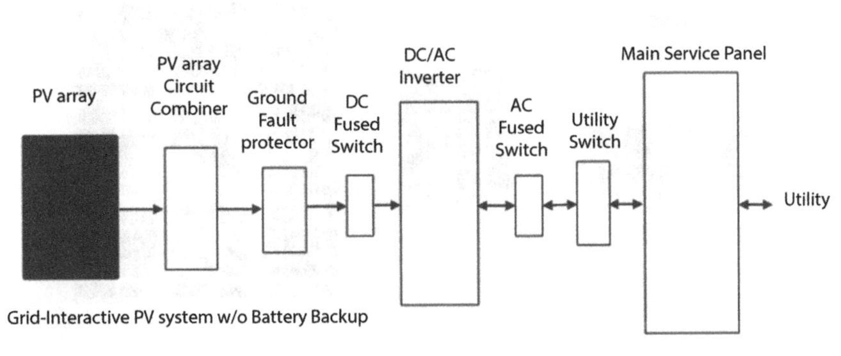

Figure 1.3 Grid-connected PV system.

generating electricity, and this power should not only be sufficient to fulfil load requirements but should also be fed back into the grid if the supply exceeds demand. Here, we examine PV modules in detail, after which we move on to inverters and batteries [25].

1.4.1 PV Module

The solar cells were then assembled into modules. When grouped together, these cells make up a PV module, which is part of a larger PV array. The features of the modules can be affected by the arrangement of the cells and modules (series and parallel). When solar cells are connected in series, their open-circuit voltages increase, resulting in a higher output voltage. However, the current flowing through the series connection remained constant under steady-state conditions. When cells are connected in parallel, the voltage across all the cells is maintained at a steady state. The current of each cell has now been added together [26].

In this article, we examine the functioning of a PV module in further detail. They have varying reactions based on their location and weather. To maximize the output, the PV modules should be exposed to bright, direct sunlight. Let us examine what it means to maximize PV module performance by adjusting the tilt and orientation of the modules. The tilt of the solar module determines the angle of elevation or pitch with respect to the horizontal plane. The azimuth of the position-related surface of the modules is defined by the degree of freedom orientation. Azimuth is defined in a variety of ways depending on the region and culture. Geographically north and south are the most commonly used coordinates [27]. Figure 1.4 shows the components of the PV array.

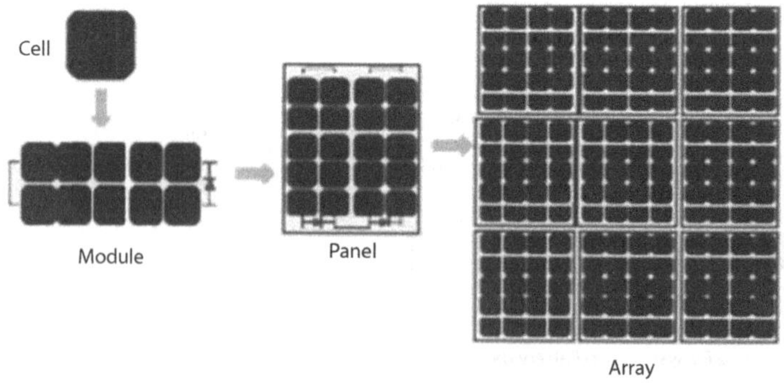

Figure 1.4 Components of the PV array.

1.4.2 PV Array and Cells

A photovoltaic (PV) array comprises interconnected solar cells that convert sunlight into electrical energy. Typically, a single solar cell produces a low voltage (typically around 0.5 to 1 V), and hence, numerous cells are linked in series to attain the desired voltage. Silicon, which is readily available and has favorable electrical characteristics, is the material of choice for most solar cells. The efficiency of a PV array is influenced by various factors, such as the number, orientation, and efficiency of the cells and shading effects. PV arrays can be configured in different ways to optimize their performance for various applications including rooftop solar panels, large-scale solar farms, and portable solar chargers.

Different types of solar cells are available, including monocrystalline, polycrystalline, thin film, and concentrator cells. Monocrystalline cells are the most efficient, but also the most expensive, because they are made from a single, pure crystal of silicon. In contrast, polycrystalline cells are less efficient, but more affordable, as they are made from multiple silicon crystals. Thin-film cells are lightweight, flexible, and suitable for various applications; however, they are inefficient. In comparison, concentrator cells are more expensive and utilize lenses or mirrors to focus sunlight onto a small area of a highly efficient PV material.

Shading can reduce the power output of a PV array because shaded cells generate less electricity than unshaded cells. This can be addressed by incorporating bypass diodes, which enable current flow around the shaded cells, or by using maximum power point tracking (MPPT) algorithms, which optimize the PV array output by regulating the voltage and current to attain the maximum power output. A typical grid-connected system involves grid-tied inverters that link the DC output of the PV array to the AC power grid, thereby contributing to renewable energy supply. Overall, PV arrays and cells are important and rapidly developing technologies in the renewable energy sector, offering a clean and sustainable source of electricity generation. As research and development continues, advancements in PV cell efficiency, material design, and array optimization are expected to further increase the economic viability and widespread adoption of solar power.

1.4.3 Solar Inverter

The solar inverter is crucial power electrical equipment that allows for the widespread use of PV systems today. It consumes the modules' DC power and makes it useful for all home electronics. Knowing the exact role that an inverter plays in a PV system is crucial for understanding what it does. Most of the country's power grids operate on the AC. Currently, most houses have air-conditioning systems. Because the current energy infrastructure is designed for use with alternating current, DC solar power must be converted to AC. Different types of inverters exist, each with its own set of advantages and disadvantages. Let us separate them accordingly to how they work. Inverters are designed specifically for use with either grid-connected or stand-alone systems, as well as bimodal inverters that can be used with either form of connection [28].

The first two types were the most common. Figure 1.3 illustrates a typical grid-connected system, where grid-tied inverters are utilized to link the DC produced by the PV modules to the AC power grid. Irradiance levels determine whether the PV system or grid is used to power the load. Therefore, inverters that feed AC power into the grid also serve as current sources, as they run at the grid voltage and frequency. If an existing power grid is available, then a grid-connected topology can be implemented. Some areas lack access to the grid, necessitating the use of a stand-alone topology. A conventional standalone PV system is shown in Figure 1.5. The inverter is a stand-alone result. Because there is no grid connection in this situation, the load is solely dependent on the PV system for power [29].

The inverter acts as a voltage source, providing either 230 V or 110 V AC at a constant voltage and frequency to the load. The extra energy from an off-grid system must be stored so that it can be used at night when no new power is produced. The inverter also serves as a maximum power point tracker, which is an additional objective. This guarantees the highest possible output from PV modules or arrays. The inverters can be broken down into distinct groups based on their implementation topology and their modes of operation [30]. The classification system included four subsets.

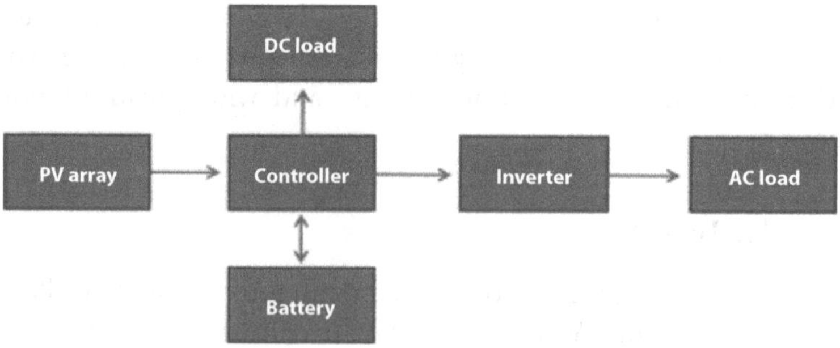

Figure 1.5 Off-grid or standalone PV system.

Central inverter (Several kW to hundred MW)
Module inverter (50 W to 500 W)
String inverter (500 W to a few kW)
Multi string inverter (1 kW to 10 kW)

1.4.3.1 *Central Inverter*

The central inverter is the most commonly used inverter topology. As shown in Figure 1.6, the implementation is straightforward, requiring only a single central inverter to power all the PV modules in the system.

The complexity of the system design is reduced owing to the inverter topology. However, there are a few limitations to this system, such as the fact that a large amount of DC power must be transmitted via long lines to get to the central inverter. Because the DC fault current is difficult to interrupt, this raises the DC wire costs and reduces safety. When integrated into a single inverter, MPPT can only serve the entire network. Because partial shading can cause modules or strings to become misaligned, the system output will drop significantly [31].

1.4.3.2 *Module Inverter*

As shown in Figure 1.7, under this inverter design, each module in a PV system has its own inverter equipped with an MPP tracker.

Therefore, this topology outperformed the central inverter topology under partial-shade conditions. Module inverters offer the highest level of system flexibility. Moreover, the DC wiring is considerably less expensive.

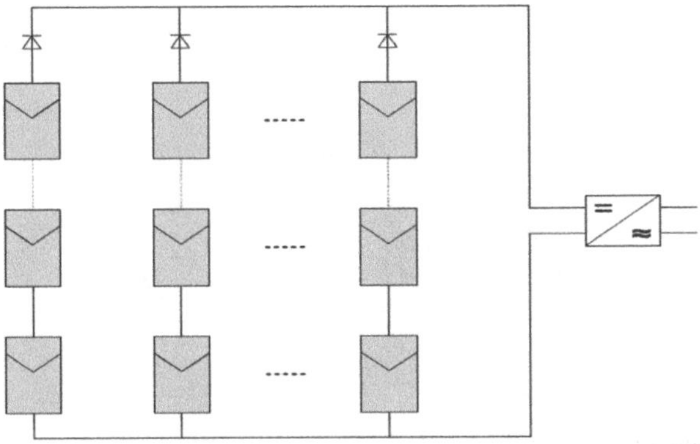

Figure 1.6 Central inverter topology.

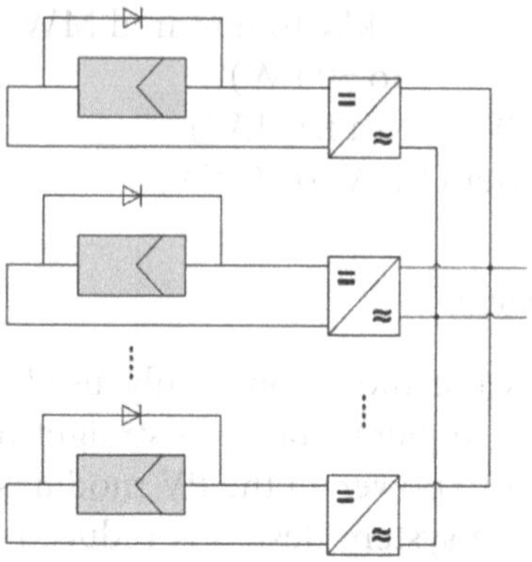

Figure 1.7 Module inverter topology.

Nonetheless, this inverter design has higher upfront costs and ongoing maintenance costs than the others [32].

1.4.3.3 String Inverter

The goal of this inverter layout is to find the middle ground between the central and module-level inverter layouts. As shown in Figure 1.8, the string inverter topology involves the provision of a dedicated inverter for each series-connected group of the PV modules. With a string inverter, each string may run at its maximum power point (MPP), making it more efficient than a central inverter in mismatch situations. Even though module inverters are simpler in theory, their implementation is more complex. Furthermore, the string inverter topology is affected more by partial shading than the module inverter topology [33].

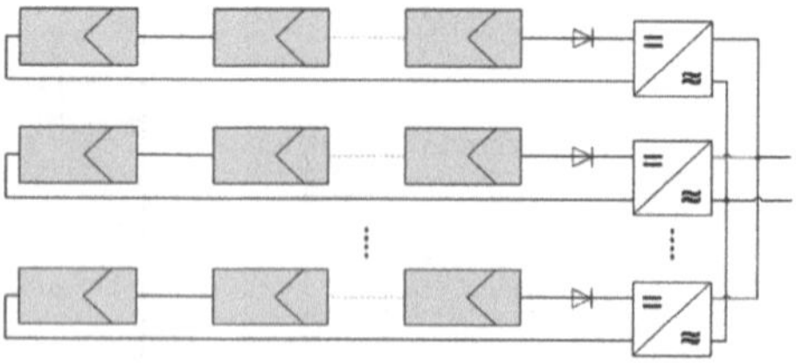

Figure 1.8 String inverter topology.

1.4.3.4 *Multi String Inverter*

This proposal is an effort to merge the advantages of the central inverter's reduced price and the high-energy-level string inverter. Low-voltage DC–DC converters are utilized to handle the power output of each string beforehand. A DC–DC converter and MPP tracker are included in each string. The DC bus connects all converters to the inverter, which is then connected to the grid [34]. Figure 1.9 shows the multi string inverter topology.

1.4.4 Characteristics of Solar Inverter

We have now completed an overview of the solar inverters and their topologies. Naturally, the requirements, scale, and cost of the system should be considered when deciding on a topology for implementation. However, the following factors also influence the selection of a solar inverter for a PV system [35].

a. This is because each component of the power-processing chain require energy to function. In general, string inverters outperform other inverter topologies in terms of their efficiency. Under the rated settings, the typical efficiency was 95%.
b. Integrated MPP trackers are a standard feature of modern inverters, although their availability varies with architecture.
c. It is anticipated that grid-connected inverters will incorporate active islanding detection features. When the grid operator cuts off electricity, the inverters in a grid-tied system continue to run; this is called islanding.
d. As solar inverters are typically placed outside, they are vulnerable to environmental factors. They must accommodate the weather and humidity levels.

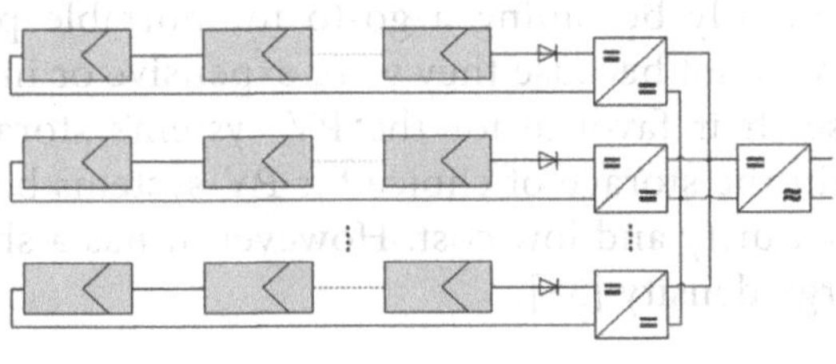

Figure 1.9 Multi string inverter topology.

e. The harmonic distortion from the solar inverters should be negligible.

1.4.5 Battery Storage in PV System

Off-grid PV systems can be a lifesavers when there is no access to an electricity grid. Battery storage is one of the most important components of an off-grid solution. There are two types of source intermittency in the context of PV systems. Changes in irradiance that occur over the course of a day and those that occur between summer and winter represent the 24-hour and seasonal variations, respectively. During the day's dark hours, the power consumption spikes significantly. When solar power is insufficient, a solitary system cannot use the excess energy produced during the day because there is nothing to put it. It is possible to store the extra solar energy produced during the day in a battery integrated into the PV system. Next, the battery can be drained during periods of low sun exposure. Thus, the supply and demand of loads can be synchronized. Many technological possibilities exist to meet storage requirements [36].

A battery is a device that converts chemical energy into electrical energy via electrochemical processes. Primary and secondary batteries are the two main categories of batteries. The chemical energy stored in the primary batteries is permanently transformed into electrical energy. Batteries based on zinc-carbon and alkalinity are examples. The chemical energy stored in the secondary batteries can be converted back into electrical power, making them rechargeable. The various battery types include lead-acid and lithium-ion batteries [37].

To this end, we are particularly interested in secondary batteries as potential energy sources. Various secondary battery technologies are already available and can be implemented. For instance, lead-acid batteries are the oldest and most developed battery technologies that are currently available. In addition, lithium-ion batteries and ion-polymer batteries are available. These are currently the subjects of extensive studies as potential storage solutions. Owing to their high energy density, they are rapidly becoming a go-to for portable power devices. However, this was not because they were expensive or immature. They did not choose their favorite for the PV system's storage. Lead-acid batteries remain the storage of choice for PV systems because of their unparalleled maturity and low cost. However, it has a shorter lifespan and lower energy density [38].

1.5 Power Losses in PV System

Power losses, such as MPP tracking losses, mismatch losses, and misrepresentative losses, are illustrated in Figure 1.10 for the PV power generation schemes. When every PV module in an array operates at the same insolation level, MPP tracking losses occur. Rapid changes in the insolation conditions often prevent maximum power point tracking (MPPT) methods from quickly reaching the maximum power point (MPP) of a PV array, resulting in losses.

Using the power–voltage characteristics of a PV system, as depicted in Figure 1.11, an MPPT scheme is tasked with performing a certain action. The power–voltage characteristics reveal the voltage at which the highest power is produced. We refer to it as "MPP voltage." As the sun's rays move

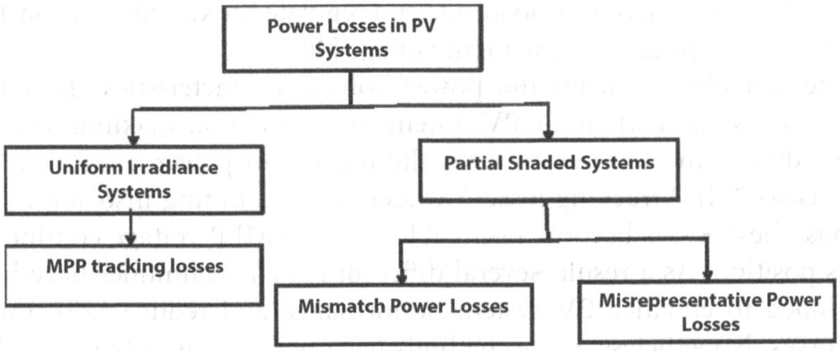

Figure 1.10 Types of the most common power losses in PV systems.

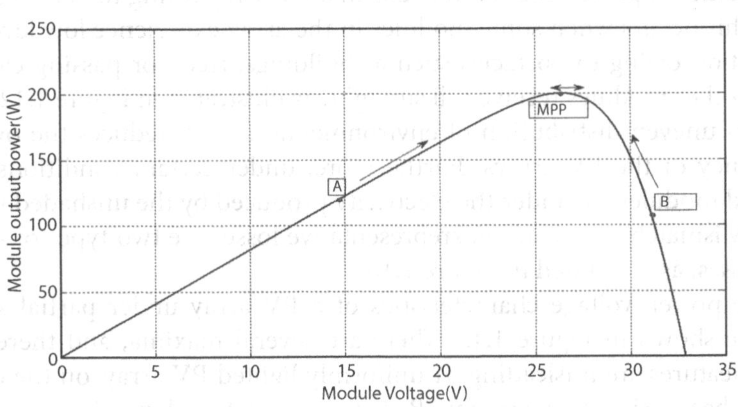

Figure 1.11 P-V characteristics showing MPP and operating points A and B.

during the day, so does the MPP voltage starting point. By ensuring that the PV array continues to operate at this varying MPP voltage, MPPT methods can be used to extract the maximum instantaneous power. To enhance the power output, the algorithm frequently uses a hill-climbing approach that involves adjusting the operational voltage [39].

The primary method involves repeatedly switching the output voltage of the PV array and analyzing the difference between the array's current power output, P (N), and its prior power output, P (N-1), to establish the direction of the subsequent voltage transition. If the PV array's power is increased because of a voltage change, the direction of the subsequent voltage change should be kept the same. The algorithm typically employs a hill-climbing technique to enhance power generation by adjusting the working voltage until it reaches and stabilizes near the maximum power point (MPP). At this point, the system can produce maximum output. To ensure that the output voltage remains constant, DC–DC converters isolate the PV system from the load side and regulate the voltage at a constant load voltage, regardless of the input voltage [40].

One can observe from the power–voltage characteristics show that power losses occur when the PV system operates at non-maximum power points during the journey towards the maximum power point. We call these costs "MPP tracking losses." Under rapidly shifting insolation conditions, these losses become noticeable, and the MPP voltage continually shifts position. As a result, several different MPPT techniques have been developed to enhance PV system performance and reduce MPP tracking losses. Nonetheless, these methods have not been widely pursued or implemented because they require precise temperature estimates, which increase the difficulty and cost of their implementation [41].

Additional power losses can occur in a PV array owing to partial shading. This occurs when some modules in the array experience low levels of insolation owing to obstacles such as buildings, trees, or passing clouds, while other modules receive full sunlight, as illustrated in Figure 1.12.

This uneven distribution of environmental factors reduces the overall efficiency of the PV arrays. Furthermore, under certain conditions, the shaded modules squander the electricity produced by the unshaded modules. Mismatch losses and misrepresentative losses are two types of shading losses, as illustrated in Figure 1.10.

The power–voltage characteristics of a PV array under partial shading are shown in Figure 1.13. There are several maxima, and therefore, these features are misleading. A uniformly lighted PV array, on the other hand, has a single maximum. Because conventional maximum power point tracking (MPPT) systems can become trapped in any of the several

Figure 1.12 An example of partially shaded PV array.

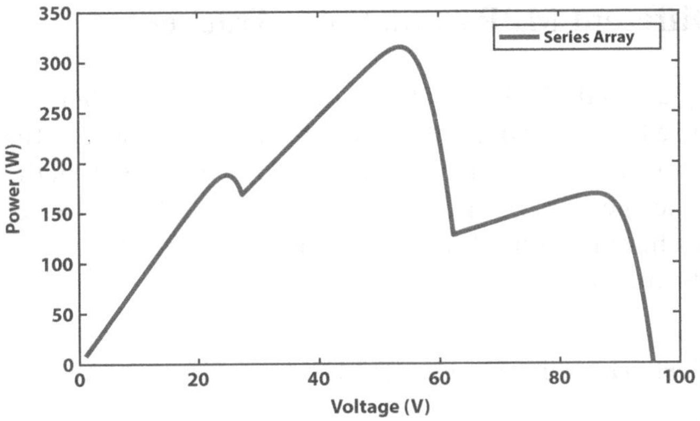

Figure 1.13 Power–voltage characteristics of partial shaded PV array.

local maxima, rather than advancing toward global maxima, MPPT is a challenging task. Misrepresentative losses occur when the available power (global maxima) exceeds the operating power (local maxima) [42].

To maximize the power output of PV arrays in partially shaded conditions, several global maximum power point tracking (GMPPT) systems have been proposed. Despite its tight coupling to the GMPP, the power–voltage characteristics occasionally require filtering, forcing the PV array to operate at less-than-optimal power points and wasting energy in the process. When the lighting conditions or shading patterns change, a new GMPP is created, and the system can begin working on this new GMPP after it has scanned its characteristics again [43].

The current–voltage curves of the modules in a PV array are not always equal, even when the modules have the same power rating and are manufactured by the same company. Any given current–voltage curve can be described by four values: Pm, Isc, Vm, and Voc. Mismatch losses in a PV system occur when the characteristics of individual modules vary. The mismatch of the current–voltage curves in PV arrays can be attributed to several issues, the most prevalent of which are a decrease in photon-generated power, erratic external soiling, manufacturer tolerance, and surface fissures. However, mismatch losses are most commonly caused by partial shading or non-uniform insolation, which leads to inconsistencies in the current–voltage curves of individual modules. Scientists and electrical engineers have spent the last decade focusing on lowering these losses [44].

1.6 Different MPPT and Solar Tracker

A technique called Maximum Power Point Tracking (MPPT) is employed to maximize the power output of photovoltaic (PV) systems. This involves the use of an algorithm to track the maximum power point of the PV array and regulate the voltage and current levels in real-time. Several MPPT techniques have been used in PV systems, each with its own advantages and disadvantages.

1.6.1 Perturb and Observe (P&O) Algorithm

The MPPT technique involves modifying the operating point of the PV system and monitoring the resulting changes in power. The algorithm continues to adjust the operating point in the same direction if the power increases, until it no longer increases. This approach can cause fluctuations around the maximum power point, which can decrease the efficiency of the system.

1.6.2 Incremental Conductance Algorithm

The incremental conductance algorithm maintains a slope of zero by tracking the power–voltage (P–V) curve of the PV array and adjusting the operating point of the PV system. It has the advantage of being able to track changes in irradiance and temperature, leading to an improved system efficiency. However, this algorithm is more complex than the P&O algorithm is.

1.6.3 Fractional Short-Circuit Current (FSCC) Algorithm

This algorithm uses the ratio of the short-circuit current to open-circuit voltage to determine the maximum power point of the PV system. This is advantageous in that it can operate even under low irradiance and is less affected by shading. However, obtaining accurate measurements of short-circuit current and open-circuit voltage can be challenging, which is a disadvantage of this algorithm.

1.6.4 Artificial Intelligence (AI) Algorithms

MPPT systems have utilized artificial intelligence (AI) algorithms such as fuzzy logic and neural networks to improve the accuracy of tracking the maximum power point. By incorporating various factors that affect the performance of the PV array, such as shading, temperature, and varying load conditions, these algorithms enhance tracking ability. However, these algorithms require more computational resources and may be more challenging to implement.

In summary, the selection of an MPPT algorithm depends on several factors, including system design, environmental conditions, and cost. The P&O algorithm is the most widely used because of its simplicity, while more intricate algorithms such as the Incremental Conductance and FSCC may offer higher efficiency. AI algorithms, while providing better accuracy and adaptability, require more computational resources and can be expensive to implement.

Solar trackers are devices that follow the sun's movement throughout the day to optimize the amount of sunlight reaching the solar panel or array. Compared with fixed-tilt solar panels, solar trackers can increase energy production by up to 40%. There are two main types of solar trackers: single-axis and dual-axis trackers. A single-axis tracker follows the sun's movement horizontally, whereas a dual-axis tracker adjusts the tilt of the solar panel to follow the sun's elevation. Solar trackers are often used in large-scale solar power plants to reduce energy costs and land requirements; however, they are generally less cost-effective in residential and commercial settings because of their higher installation and maintenance costs. However, there are challenges associated with solar trackers. They require a more complex and expensive installation process than fixed-tilt solar panels, and require more maintenance to sustain proper functioning. In addition, solar trackers may be less effective in areas with frequent cloud cover or high winds, which can cause the tracker to move out of alignment. Overall, solar trackers are promising for maximizing the energy output of solar panels and reducing the cost of solar power. Continued development and improvement of solar tracker technology can help overcome the challenges associated with this technology and make it more accessible and cost-effective for a wide range of applications.

1.7 Development in Standalone PV System

Standalone PV systems, also known as off-grid PV systems, are designed to satisfy the electricity needs of isolated or remote areas that are not connected to the main power grid. These systems are becoming increasingly popular in rural and remote areas as they provide a reliable and sustainable source of electricity. The development of standalone PV systems has been driven by advances in PV technology and increasing demand for clean energy sources. In recent years, the efficiency and cost of PV modules have improved significantly, making them a viable option for standalone systems. In addition, the development of new battery technologies, such as lithium-ion batteries, has made it possible to store excess energy generated by PV systems for use when there is no sunlight.

Designing a reliable and efficient standalone PV system that can satisfy the energy demands of users is a significant challenge. This involves careful consideration of factors such as the system location, PV array, and battery bank size, and energy requirements of users. Furthermore, standalone PV systems must be designed to endure severe environmental conditions, such as extreme temperatures and weather events, that can affect the system's performance and lifespan.

Another challenge is the integration of standalone PV systems with the existing energy infrastructure. In many cases, standalone PV systems are designed to replace diesel generators or other fossil fuel-based power sources. However, integrating these systems with the existing infrastructure can be complex and requires careful planning to ensure compatibility and reliability.

Despite these challenges, the development of standalone PV systems has the potential to provide clean and sustainable energy for millions of people worldwide. As technology continues to improve and costs continue to decrease, standalone PV systems are likely to become an increasingly important part of the global energy mix [45].

1.8 The Development and Challenges in DC–DC Converter for PV Applications

Photovoltaic (PV) technology has gained momentum in recent years as a promising renewable energy source. However, the efficiency and stability of a PV system depend heavily on the power conditioning equipment, including the DC–DC converter. The interface between the load and PV array is essential, and the DC–DC converter has undergone significant advancements and challenges in its design and implementation.

The first generation of DC–DC converters used linear regulators, which provided a stable output voltage but incurred high losses and low efficiency. Subsequently, pulse-width modulation (PWM) and maximum power point tracking (MPPT) techniques were introduced to improve converter efficiency and performance. The MPPT technique helps optimize the power output of a PV array by tracking the maximum power point under changing environmental conditions. Advanced converter topologies, including dual-active bridges, interleaved buck-boost, and resonant converters, have been developed to improve converter efficiency and reliability. These topologies offer advantages, such as reduced component stress, better power density, and higher efficiency. Additionally, the utilization of wide-bandgap semiconductors, such as silicon carbide and gallium nitride, has led to better converter efficiency and decreased converter size and weight.

Despite these advancements, several challenges still need to be addressed in the design and implementation of DC–DC converters for PV applications. One of the key challenges is the ability of the converter to operate under varying environmental conditions, including temperature, irradiance, and shading. These conditions affect the efficiency, stability, and reliability of the converters. In addition, the converter must be designed to accommodate different PV module technologies, including crystalline silicon, thin films, and emerging perovskite cells.

Another challenge is the converter's cost and size, which are important factors that impact both the cost and performance of the PV system. The development of low-cost, high-performance converters with reduced size and weight is critical for the widespread adoption of PV technology [46].

1.8.1 Recent Development in Microinverters for PV Applications

In the field of photovoltaic (PV) systems, microinverters have gained popularity due to their numerous advantages over conventional string inverters. A microinverter is a small electronic device that converts DC power generated by each solar panel in a PV system into AC power that can be used in households or fed into the grid. Recent developments in microinverters have focused on increasing efficiency, improving reliability, and reducing costs. An advancement in the power electronics of microinverters is the use of wide bandgap (WBG) semiconductors such as silicon carbide (SiC) and gallium nitride (GaN). These semiconductors have characteristics that are superior to traditional silicon-based semiconductors such as lower power losses, smaller sizes, and higher operating temperatures, which make them a suitable option for microinverters.

Another development is the incorporation of advanced control algorithms and communication protocols in microinverters. These algorithms and protocols enable real-time monitoring of each solar panel's performance, allowing for early detection of faults and optimization of power production. Additionally, the use of modular designs in microinverters has made it easier to maintain and replace individual components, reducing downtime and maintenance costs.

One challenge in the development of microinverters for PV applications is the need to balance high efficiency and reliability with low cost. While WBG semiconductors and advanced control algorithms can improve efficiency and reliability, they also increase costs. Therefore, finding a balance between cost and performance remains a major challenge.

Another challenge is the integration of microinverters into larger PV systems. While microinverters offer benefits such as module-level power optimization and flexibility in system design, they also require additional hardware and software for communication and monitoring. Integration of microinverters with other components in a PV system requires careful consideration of compatibility, scalability, and reliability [47].

1.9 PV-Powered Electric Vehicles

PV-powered electric vehicles, also known as solar-powered electric vehicles, are a sustainable and environment-friendly solution for transportation. These vehicles are powered by solar energy captured by PV panels mounted on the roof of the vehicle or combined with the vehicle's body. PV-powered electric vehicles offer several advantages over the conventional electric vehicles. First, they can recharge their batteries using clean, renewable energy from the sun, thereby reducing their reliance on grid electricity. This makes them an ideal solution for remote areas or off-grid communities, where access to electricity is limited. Moreover, PV-powered electric vehicles can significantly reduce the carbon footprint of transportation because they emit zero greenhouse gases during operation. This can help mitigate the environmental impact of transportation, which is a major contributor to global carbon emissions.

Despite the benefits of PV-powered electric vehicles, there are several challenges in their widespread adoption. The efficiency of PV panels remains relatively low, which limits the range and speed of these vehicles. The cost of PV panels is also relatively high, which can make PV-powered electric vehicles expensive than conventional electric or gasoline-powered vehicles. Furthermore, the integration of PV panels into the vehicle's body can be challenging because it requires a redesign of the vehicle's structure and can add weight to the vehicle. This can affect vehicle performance, particularly in terms of acceleration and handling. To overcome these challenges, research and development of PV technology and vehicle design are ongoing. Advances in PV technology are improving the efficiency and reducing the cost of PV panels, which could render PV-powered electric vehicles more affordable and practical in the future.

In addition, vehicle manufacturers are exploring new ways to integrate PV panels into the vehicle body, such as flexible PV panels that can be shaped to fit the contours of the vehicle. This could make PV-powered electric vehicles more aerodynamic and improve their performance [48, 49].

1.10 Discussion

Solar photovoltaic technology has advanced significantly since its beginning in the mid-1900s. While the first solar cell was invented by Bell Labs in 1954, practical applications for solar PV systems became available only in the 1970s, such as powering satellites and remote telecommunications equipment. Throughout the 1980s and the 1990s, solar PV technology continued to develop, and government incentives for renewable energy drove industrial growth. However, it was not until the early 2000s that solar PV started to become more competitive with traditional energy sources in terms of cost. In recent years, significant developments have been made in the solar PV technology. Improvements in cell efficiency and manufacturing processes have led to a decrease in the costs of solar PV systems. Furthermore, the development of energy storage technology, including but not limited to lithium-ion batteries, has enabled the storage of surplus solar energy during periods of insufficient sunlight. This has made solar PV systems more reliable and viable for a wide range of applications.

One area of significant development in recent years is the integration of solar PV systems with electricity grids. The growth of distributed generation, where electricity is generated and used locally, has driven the emergence of novel technologies, such as microinverters and intelligent inverters. These devices enhance the effectiveness of solar PV systems and enable them to be integrated more easily into the grid. Another innovative advancement in the solar photovoltaic technology is the implementation of bifacial solar panels that can efficiently produce electricity from both the front and rear sides of the panel. This has the potential to increase the energy yield of solar PV systems and increase their efficiency.

Additionally, the growth of the electric vehicle market has created new opportunities for solar PV systems. PV-powered electric vehicles can be charged using solar panels, making them more sustainable and reducing their reliance on traditional energy sources. Overall, the history of solar PV technology has been characterized by steady growth and development, with significant advancements in recent years. As the global community endeavors to shift towards a more environmentally sustainable future, solar PV technology is expected to play a more significant role in satisfying our energy requirements.

1.11 Conclusion

Solar photovoltaic (PV) systems play an essential role in compensating for electrical energy shortages. The reason for this is that there is a lack of electricity owing to rising demand and the decreasing use of traditional energy sources. The maximum amount of power that can be generated by the PV array is reduced when it is partially shaded. This chapter provides a review of the trends in PV systems, a historical overview, and grid and off-grid PV systems. This chapter plays a vital role in gaining knowledge on PV systems.

References

1. International Energy Agency, "Global energy demand rose by 2.3% in 2018, its fastest pace in the last decade, news," *2019-03-26*, 2019. https://www.iea.org/newsroom/news/2019/march/global-energy-demand-rose-by-23-in2018-its-fastest-pace-in-the-last-decade.html
2. M. F. Jalil, S. Khatoon, I. Nasiruddin & R. C. Bansal, "Review of PV array modelling, configuration and MPPT techniques", *Int. J. Model. Simul.*, 42:4, 533-550, 2022, DOI: 10.1080/02286203.2021.1938810.
3. G. Meerimatha and B. L. Rao, "Novel reconfiguration approach to reduce line losses of the photovoltaic array under various shading conditions," *Energy, 196*, 117120, 2020.

4. M.F. Jalil, M. S. Ansari, S. Diwania, M. A. Husain, "Performance Analysis of PV Array Connection Schemes Under Mismatch Scenarios", *Renewable Power for Sustainable Growth, Lecture Notes in Electrical Engineering*, vol. 723, 2021, Springer, Singapore, https://doi.org/10.1007/978-981-33-4080-0_22.
5. International Energy Agency, "Renewables Information 2018: Overview, Report," 2018. https://www.iea.org/reports/renewables-information-overview/transformation
6. International Energy Agency, "Trends 2018 in Photovoltaics Applications, Report," 2018. https://iea-pvps.org/wp-content/uploads/2020/01/2018_iea-pvps_report_2018.pdf
7. International Energy Agency, "Market Report: Renewables 2018," 2018. doi: 10.1787/coal_mar-2017-en. https://www.iea.org/reports/renewables-2018
8. S. Khatoon, Ibraheem, and M. F. Jalil, "Analysis of solar photovoltaic array under partial shading conditions for different array configrations," in *2014 Innovative Applications of Computational Intelligence on Power, Energy and Controls with their impact on Humanity (CIPECH)*, Nov. 2014, pp. 452–456, doi: 10.1109/CIPECH.2014.7019127.
9. S. Husain and N. A. Shrivastava, "A Comparative Analysis of Multi-objective Optimization Algorithms for Stand-Alone Hybrid Renewable Energy System," *2020 2nd International Conference on Innovative Mechanisms for Industry Applications (ICIMIA)*, 2020, pp. 255-260.
10. M. F. Jalil, S. Khatoon, I. Nasiruddin, and R. C. Bansal, "An Improved Feasibility Analysis of Photovoltaic Array Configurations and Reconfiguration Under Partial Shading Conditions," *Electr. Power Components Syst.*, vol. 48, no. 9–10, pp. 1077–1089, 2020.
11. S. R. Potnuru, D. Pattabiraman, S. I. Ganesan, and N. Chilakapati, "Positioning of pv panels for reduction in line losses and mismatch losses in pv array," *Renewable Energy*, vol. 78, pp. 264–275, 2015.
12. International Energy Agency, "Solar PV, Tracking Clean Energy Progress, Report," 2019. https://www.iea.org/reports/tracking-sdg7-the-energy-progress-report-2019
13. Renewable Energy Policy Network, "Global status report 2018," 2018. https://www.ren21.net/wp-content/uploads/2019/05/GSR2018_Full-Report_English.pdf
14. Ministry of new and renewable energy, "Annual report 2017-18," 2018. https://mnre.gov.in/annual-reports-2017-18/
15. A. E. Becquerel, "Mémoire sur les effets électriques produits sous l'influence des rayons solaires," *Comptes Rendus des Séances Hebdomadaires*, vol. 9, pp. 561–567, 1839, doi: 10.1016/j.jpedsurg.2007.08.042.
16. W. G. Adam and Day.R.E, "The action of light on selenium," *Philos. Trans. R. Soc. London*, vol. 167, pp. 313–349, Jan. 1877, doi: 10.1098/rstl.1877.0009.
17. C. E. Fritts, "New Form of Selenium Cell," *Am. J. Sci.*, vol. 3–26, no. 156, pp. 465–472, 1883, doi: 10.2475/ajs.s3-26.156.465.

18. M. Planck, "On an Improvement of Wien's Equation for the Spectrum," *Annalender Phyisk*, vol. 2, no. 202, pp. 79–81, 1900, doi: 10.1016/B978-0-08-012102-4.50012-7.
19. Grondahl. L. O, "The Copper-Cuprous-Oxide Rectifier and Photoelectric Cell," *Rev. Mod. Phys.*, vol. 5, pp. 141–168, 1933.
20. R. S. Ohl, "Light-sensitive electric device including silicon, Patent," 1941.
21. D. M. Chapin, C. S. Fuller, and G. L. Pearson, "A New Silicon p-n Junction Photocell for Converting Solar Radiation into Electrical Power," *J. Appl. Phys.*, vol. 25, no. 5, pp. 676–677, May 1954, doi: 10.1063/1.1721711.
22. Dr. S. Philips and W. Warmuth, "Fraunhofer ISE: Photovoltaics Report," 2019. https://www.ise.fraunhofer.de/content/dam/ise/en/documents/annual_reports/fraunhofer-ise-annual-report-2019-2020.pdf
23. M. A. Green *et al.*, "Solar cell efficiency tables (Version 53)," *Prog. Photovoltaics Res. Appl.*, vol. 27, no. 1, pp. 3–12, 2019, doi: 10.1002/pip.3102.
24. B. Yang, H. Ye, J. Wang, J. Li, S. Wu, Y. Li, H. Shu, Y. Ren, H. Ye, "PV arrays reconfiguration for partial shading mitigation: Recent advances, challenges and perspectives," *Energy Convers. Manage.*, volume 247, 2021, 114738, ISSN 0196-8904, https://doi.org/10.1016/j.enconman.2021.114738.
25. M. Azharuddin, T. S. Babu, N. Bilakanti, N. Rajasekar, "A nearly accurate solar photovoltaic emulator using a dspace controller for real-time control", *Electric Power Components and Systems*, 44 (2016) 774–782.
26. N. Rajasekar, N. K. Kumar, R. Venugopalan, "Bacterial foraging algorithm based solar pv parameter estimation," *Sol. Energy*, 97 (2013) 255–265.
27. T. S. Babu, J. P. Ram, K. Sangeetha, A. Laudani, N. Rajasekar, "Parameter extraction of two diode solar pv model using fireworks algorithm," *Sol. Energy*, 140 (2016) 265–276.
28. D. Alam, D. Yousri, M. Eteiba, "Flower pollination algorithm based solar pv parameter estimation," *Energy Convers. Manage.*, 101 (2015) 410–422.
29. B. Jacob, K. Balasubramanian, S. M. Azharuddin, N. Rajasekar, *et al.*, "Solar pv modelling and parameter extraction using artificial immune system," *Energy Procedia*, 75 (2015) 331–336.
30. B. Jacob, K. Balasubramanian, T. S. Babu, N. Rajasekar, "Parameter extraction of solar pv double diode model using artificial immune system," in: *2015 IEEE International Conference on Signal Processing, Informatics, Communication and Energy Systems (SPICES)*, IEEE, pp. 1–5.
31. D. Yousri, T. S. Babu, D. Allam, V. K. Ramachandaramurthy, M. Eteiba, "Fractional chaotic ensemble particle swarm optimizer for identifying the single, double, and three diode photovoltaic models' parameters," *Energy*, 2020, 116979.
32. D. Allam, D. Yousri, M. Eteiba, "Parameters extraction of the three diode model for the multicrystalline solar cell/module using moth-flame optimization algorithm," *Energy Convers. Manage.*, 123, 2016, 535–548.
33. S. Harb, M. Kedia, H. Zhang and R. S. Balog, "Microinverter and string inverter grid-connected photovoltaic system — A comprehensive study," *2013*

IEEE 39th Photovoltaic Specialists Conference (PVSC), 2013, pp. 28852890, doi: 10.1109/PVSC.2013.6745072.

34. N. A. Rahim and J. Selvaraj, "Multistring Five-Level Inverter With Novel PWM Control Scheme for PV Application," in *IEEE Trans. Ind. Electron.*, vol. 57, no. 6, pp. 2111-2123, June 2010, doi: 10.1109/ TIE.2009.2034683.
35. N. Kim and B. Parkhideh, "PV-Battery Series Inverter Architecture: A Solar Inverter for Seamless Battery Integration With Partial-Power DC– DC Optimizer," in *IEEE Trans. Energy Convers.*, vol. 34, no. 1, pp. 478-485, March 2019, doi: 10.1109/TEC.2018.2873664.
36. J. Tant, F. Geth, D. Six, P. Tant and J. Driesen, "Multiobjective Battery Storage to Improve PV Integration in Residential Distribution Grids," in *IEEE Trans. Sustainable Energy*, vol. 4, no. 1, pp. 182-191, Jan. 2013, doi: 10.1109/ TSTE.2012.2211387.
37. B. Zhang, P. Dehghanian and M. Kezunovic, "Optimal Allocation of PV Generation and Battery Storage for Enhanced Resilience," in *IEEE Trans. Smart Grid*, vol. 10, no. 1, pp. 535-545, Jan. 2019, doi: 10.1109/ TSG.2017.2747136.
38. R. Khezri, A. Mahmoudi and M. H. Haque, "Optimal Capacity of Solar PV and Battery Storage for Australian Grid-Connected Households," in *IEEE Trans. Ind. Appl.*, vol. 56, no. 5, pp. 5319-5329, Sept.-Oct. 2020, doi: 10.1109/ TIA.2020.2998668.
39. T. -F. Wu, C. -H. Chang, L. -C. Lin and C. -L. Kuo, "Power Loss Comparison of Single- and Two-Stage Grid-Connected Photovoltaic Systems," in *IEEE Trans. Energy Convers.*, vol. 26, no. 2, pp. 707-715, June 2011, doi: 10.1109/ TEC.2011.2123897.
40. A. Chouder & S. Silvestre, "Automatic supervision and fault detection of PV systems based on power losses analysis", *Energy Convers. Manage.*, 51(10), 1929-1937, 2010.
41. A. Chouder, & S. Silvestre, "Analysis model of mismatch power losses in PV systems," *J. Sol. Energy Eng.*, 2009.
42. S. P. Jaiswal, V. Shrivastava, & D. K. Palwalia, "Opportunities and challenges of PV technology in power system", *Mater. Today.: Proc.*, 34, 593597, 2021.
43. S. R. Pendem, & S. Mikkili, "Modelling and performance assessment of PV array topologies under partial shading conditions to mitigate the mismatching power losses", *Sol. Energy*, 160, 303-321, 2018.
44. B. Dhanalakshmi, & N. Rajasekar, "Dominance square based array reconfiguration scheme for power loss reduction in solar PhotoVoltaic (PV) systems", *Energy Convers. Manage.*, 156, 84-102, 2018.
45. M. A. Husain, A. Jain, and A. Tariq, "A novel fast mutable duty (FMD) MPPT technique for solar PV system with reduced searching area," *J. Renewable Sustainable Energy*, 8(5), 054703, 2016.

46. M. Naseem, M. A. Husain, A. F. Minai, A. N. Khan, M. Amir, J. Dinesh Kumar, and A. Iqbal, "Assessment of meta-heuristic and classical methods for GMPPT of PV system," *Trans. Electr. Electron. Mater.*, 22, 217-234, 2021.
47. M. A. Husain, A. Tariq, S. Hameed, M. S. B. Arif, and A. Jain, "Comparative assessment of maximum power point tracking procedures for photovoltaic systems," *Green Energy Environ.*, 2(1), 5-17, 2017.
48. M. A. Husain, A. Jain, A. Tariq, and A. Iqbal, "Fast and precise global maximum power point tracking techniques for photovoltaic system," *IET Renewable Power Gener.*, 13(14), 2569-2579, 2019.
49. M. Naseem, M. A. Husain, J. D. Kumar, A. F. Minai, A. Ahmad, S. M. Ali, and A. S. Ansari, "A Spider Monkey Optimization Based Global Maximum Power Point Tracking Technique for Photovoltaic Systems," In *2022 2nd International Conference on Emerging Frontiers in Electrical and Electronic Technologies (ICEFEET)* (pp. 1-6), IEEE, 2022.

2

Evolution and Modeling of Solar Photovoltaic Cells: From Early to Modern Concepts

Mohammad Shahabuddin[1], Mohammed Asim[1*] and Adil Sarwar[2]

[1]Department of Electrical Engineering, Integral University, Lucknow, Uttar Pradesh, India

[2]Department of Electrical Engineering, Aligarh Muslim University, Aligarh, Uttar Pradesh, India

Abstract

The solar energy had been utilized even before 4000 B.C. for the development of mankind, it is not a new invention. In ancient times the people who were residing in rural area were building their homes with opening in the south in order to maximize the indoor temperature during low winter seasons. In 200 B.C. the concave mirror was discovered by the Chinese and later by the Greeks that points solar energy on a flammable substance to generate fire. It was used for different purposes like: cooking, soldering, and distilling perfumes. During Ist Century A.D. solar heat trapping was developed and Romans were using windows with transparent glass which allow sunlight and holds it in the head. The first solar engine was developed in 1615. Solomon De Caux invented the first solar engine. Glass lenses concentrate solar irradiation to an airtight metal vessel containing water and air. As the vessel heats up, it creates a small water fountain. This was the first published documentation of solar power use after the fall of Rome. The first solar cell was introduced by Charles Fritts and was manufactured from selenium in 1883. In April 1954, researchers present the first actual Silicon Solar PV cell at Bell Laboratory. It converts irradiation into electrical current, but past solar PV cells were having poor efficiency employed frequently. The history of solar PV cells reaches the original vision of the photovoltaic effect. In 1839, French physicist A. E. Becquerel, son of naturalist A. C. Becquerel and father of physicist H. Becquerel, was doing experiment with metal electrodes on an electrolyte solution when he

**Corresponding author*: masim@iul.ac.in

Mohammed Aslam Husain, Md Waseem Ahmad, Farhad Ilahi Bakhsh, P. Sanjeevikumar and Hasmat Malik (eds.) Photovoltaic Systems Technology, (27–42) © 2024 Scrivener Publishing LLC

saw it, he said small electromagnetic radiation are produced when substances faces to light, but the Phenomenon was not explained. It was mentioned in The New York Times that the Silicon Solar PV cell "may indicate the starting of a new span, one that will accidently lead to the realization of one of mankind's most cherished dreams — the use of uncanny power for solar energy. In past silicon solar PV cells were costly to be manufactured, and initial marketing efforts were not very successful at first. But within a short time, solar PV cells were widely used to power satellites, and other applications.

Keywords: Solar cell, fossil fuel, renewable energy, efficiency, photovoltaic module

2.1 Introduction

Horace-Benedict de Saussurea, a Swiss scientist, formulated a solar cell collector in 1767, which is supposed to be the beginning of the solar cell era. He fabricated an enclosed box in the company on one side open and placed three glass coatings on it; when heat (solar radiation) was passed through it, he observed that it amplified this radiation up to 230 °F. Subsequently, this was used in several applications. In 1953, a study on climate change was published. Both Time and popular media have introduced articles on this topic. Researchers have begun to realize the depletion of fossil fuels. Therefore, more renewable energy substances are needed. The oldest experiment a solar PV cell was developed at Bell Labs by D. Chaplin, G. Pearson, and Calvin S. F. in 1954. After some time, US Signal Corps Laboratories invented solar PV cells for satellites (earth-orbiting) to guide the solar PV panel for the space mission of Vanguard1. With further developments, solar energy has become an accepted energy source for space applications [1].

About two centuries ago, in 1839, photovoltaic effect was observed by Edmonds Becquerel when he was working in his father's laboratory at the age of 18 years. He found that when light falls on one of the electrodes immersed in a conducting solution, there is an increase in the potential difference. This is shown in the layout of the equipment explained by Edmonds Becquerel in Figure 2.1, which clearly indicates that two electrodes of a conducting material (pt) are immersed in an acid solution and when light falls on an electrode, it is observed that the voltage is enhanced.

The photovoltaic graph has fallen in terms of cost during the last few decades, and its employment as electrical energy is increasing in a moderate proportion yearly. The government is also taking an interest for its expansion and providing reasonable subsidies for its installation [2].

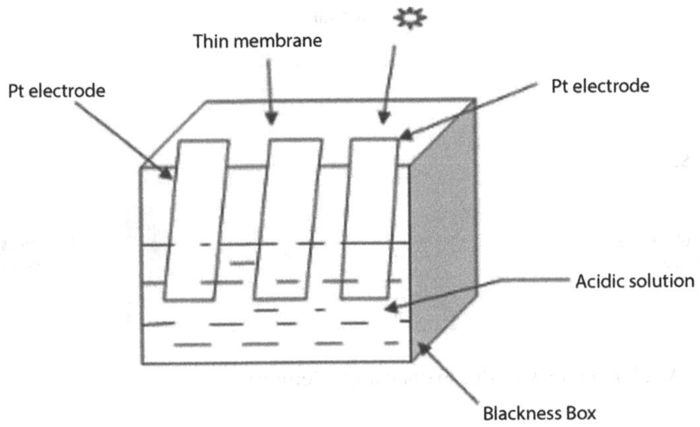

Figure 2.1 Layout of equipment explained by Edmonds Becquerel [4].

To achieve an efficient PV system, modeling is necessary for a complete study that involves basic equations for a good model [3].

This chapter is organized as follows: In Section 2.1, an introduction is specified, which describes a study on climate change. In Section 2.2, the history of solar cells is given, which illustrates the invention of PV cell technology from the work of a French physicist. In Section 2.3, the modeling of a solar PV cell is presented. In Section 2.4, the formation of a solar PV cell is summarized, and finally Section 2.5 ends with the applications of solar PV cells.

2.2 History of Solar Cell

Swiss scientist Horace-Benedict de Saussure formulated a solar cell collector in 1767, which is assumed to be the beginning of the solar cell era. Becquerel distinguished the photovoltaic in 1839 effect through a terminal in a conductive liquid manifest to light [4]. Day and Adams investigated the PV effect in selenium and wrote an article on this cell in 1877 [5]. The solar panel shown in Figure 2.2, was invented by Charles Fritts in 1883 by covering it with a thin layer of selenium and an extremely thin layer of gold. The resulting battery exhibited an electrical conversion efficiency of approximately 1%. According to Fritz, when light strikes the box, the electrons pass through the selenium and create an electric current, which is then sent through a wire to one end of the box: a direct and constant current of intensity. This invention led to the movement of solar energy production. Hallwachs developed copper and copper oxide-based solar PV

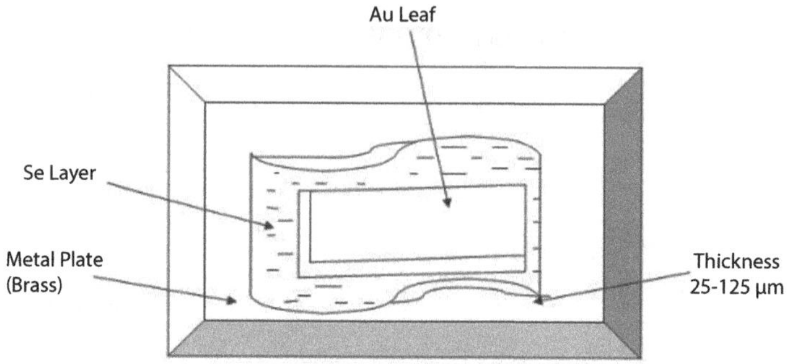

Figure 2.2 A solar cell by Charles Fritts using selenium [10].

cells in 1904. Einstein introduced an article elaborating on the PV effect on a number basis in 1905 [6]. Czochralski, creates a process to develop single crystals of metal. A few years after this process, single-crystal silicon was redesigned in 1918. Bloch introduced a new theory formed on single crystals in 1928 [7]. Wilson introduced the semiconductor process with excessive purity in 1931 [8]. Little and Teal remodeled the Czochralski process of crystal germination to produce single-crystalline germanium and, later, silicon in 1948 [9].

From 1839 to 1904, several discoveries were made in terms of the photovoltaic effect. After Becquerel observed the photovoltaic effect on an electrode immersed in an electrolyte, this effect was observed by Adams and Day in 1877 in solidified selenium and published a paper on the selenium cell. In 1883, Fritts invented a solar cell using selenium on a thin layer of gold to form a device with less than 1% efficiency. In 1904, Hallwachs fabricated a PV solar cell containing copper and copper oxide. Mohammad M. Atalla, an Egyptian engineer, invented the methods of silicon surface metallization in 1957 at Bell Laboratories by thermal oxidation. This is critical for cell efficiency. Solar cells were finally used in the vanguard satellite in 1958, where they were used as an additional power source to primary batteries to enhance the mission. They were used on the outer surface of the satellite. In 1959, US Satellite Explorer employed solar systems. In 1960 and onward, the application of solar cells became a prominent part of space satellite programs. Although the cost of solar cells did not decrease owing to the demand of space users for a high class of solar cells, the improvement in manufacturing was still gradual. Research becomes prominent for terrestrial applications of solar cell for that an eight-year program (1969 to 1977) where launched named "Research applied to natural needs" and huge finding were done for the development of solar power for ground electrical power system. Due to the oil crisis in 1973, oil companies started to purchase solar module manufacturing companies and because larger producers of solar PV module manufacturing companies of solar PV modules, various technology companies such as General Electric Motor, IBM, etc., also participated.

The first practical solar PV cell was manufactured at Bell Laboratories, and it was publicly shown, the three people named: Calvin S.F., Daril Chapin and Gerald Pearson were the inventor in 1954.

With the use of conventional energy sources such as coal, diesel, gas, the temperature is increasing daily, and carbon dioxide emissions create pollution, which is harmful for the civilization of the earth. The problem can be overcome with the use of renewable energy sources, particularly solar PV energy. It is planned to reduce the Earth's temperature by 1.5° by incorporating 50% renewable energy sources by 2050, bringing carbon dioxide emissions to zero. It is not long back but recently amazing results of silicon solar cell in terms of the efficiency are obtained, a certified efficiency of 21.9% was obtained [10].

2.3 Solar PV Cell Formation

The development of solar PV cells involves various processes. Silicon is the second most abundant substance on the surface of earth and has a unique portrait in the formation of solar PV cells. Although a few other materials that have reasonable solar PV cell formation of solar PV cell passes through various processes, silicon is the second most abundant material on the earth's crust and presents a significant part in the production of PV cells, all of which have reasonable efficiency and have efficiency of more than 40%, although they are not commercially available owing to high cost. Silicon solar PV cells are still in use and are commercially available. Solar PV cells are have monocrystalline, polycrystalline, and amorphous structures. In the process of forming silicon solar PV cells, sand is processed with carbon at 2,000 °C and 99.9% silicon is extracted from sand; then, it is converted in gaseous form and processed with hydrogen, and 98.8% silicon cannot produce current when light strikes the silicon crystal. When the load is connected, there is no flow of current because when photons strike, electrons do not move in a particular direction but move randomly. When a pentavalent impurity such as phosphorus arsenic at 17 is added to a pure silicon crystal, a large number of free electrons become available, and it is called an n-type silicon semiconductor. When a trivalent impurity such as boron is injected into a pure silicon crystal, a large number of holes become available and is called p-type silicon. In the case of a solar PV cell, the case is slightly different from an ordinary PN junction diode in the formation of a solar PV cell is a highly doped thin layer of n-type silicon brought in contact with a lightly doped wide layer of p-type silicon semiconductor. At room temperature, the free electrons from n-type silicon migrate towards the p-type silicon and combine with the whole in the p-type semiconductor; thus, a positive charge is developed at the n-type, and a negative charge is developed on the p-type side. Thus, a potential barrier is formed, and this potential barrier is formed in depletion layer, this potential barrier voltage depends on the type of material used and temperature and doping. It is 0.3 V in case of germanium and 0.7 V in case of silicon. The depletion layers play an important role in the current generation of solar PV cells. When light strikes the thin n-type layer and penetrates the depletion layer, electron-hole pairs are generated, and the current starts to flow through the load.

Previously, only single-junction silicon solar PV cells with an efficiency of 15%–20% have been discussed. However, in tandem solar cells, which incorporate multijunction crystalline silicon structures, can the efficiency be improved in the range of 30%–40% by receiving different solar spectra, such as blue and infrared, by a higher band junction on the top of the low band junction? Examples of tandem solar cells that provide higher efficiency are in the bulk category, GaAsGaSb, which has an efficiency of 34.2%, GaAlAs/Ga As with an efficiency of 27.6%, and in the thin films category CLEFT, GaAs/CIS with an efficiency of 25.8%, and a-Si/CIS with an efficiency of 14.6% [11].

Continuous improvements have been achieved in PV cell manufacturing during the last few years in various projects; Table 2.1 shows the growth in PV cell manufacturing and RND during the last few years for a significant portion of the PV manufacturing industry. There was a significant increase in the number of growers and sheets/ingots, although the wafer thickness decreased [12].

In 2014, there was a growth of 17% in PV industry shipments. From 1950 to 2000, there has been an increase in the price of all PV technologies, and there is a shortage in crystalline technology, whether it is mono crystalline or polycrystalline, and there is a growth in thin film technologies. The excitement created with the entry of amorphous silicon (a-Si) was expected to be less expensive than crystalline based PV technologies (mono- and polycrystalline). In crystalline shipment, China and Taiwan were leaders, and in thin-film shipment Malaysia and Japan, the leader [13].

Table 2.1 Optimized improvement of PV cell [14].

Advancement in R&D	Increase percentage average
Increased size of cell	50%
Increase in the number in growers	202%
Output improvements	192%
Wafer thickness reduction	25%
Increased sheet/ingot	122%
Performance increment	15%

The first of a variety of solar cells to power range machines was deployed on the Forward I Area (Point 5) satellite in 1958 and operated up to its retirement in 1964. The remarkable performance of PV cells in providing a myriad of pollution-free energy has attracted the attention of these solar PV cells. Despite their performance and price, solar PV cells are now a widely employed power source in the aerospace industry owing to their cleanness and limited energy sources. Since its discovery in France in the early 19th century, its use has led to improvements in solar cells [15].

The increasing demand for electrical power and depletion of available standard energy resources have attracted researchers' attention to the invention of renewable energy sources for generating electrical energy, such as solar energy, wind power, and seawater movement. Among these renewable energy sources, solar energy is more or less evenly available in most parts of the world and has great potential in the global energy market, including green power generation, low maintenance, and most importantly, free of charge. They are regarded as major players in solar power generation. It can generate electrical energy by directly converting sunlight into direct current or solar thermal energy. Photovoltaic energy has skyrocketed over the past decade as the Indian government has invested a lot of money into its fruitful use. Electrical energy can be obtained from PV modules or arrays in the form of direct current on sunny days, converting it to alternating current using inverters, stored using batteries, and used at night or during periods when solar radiation is not available during the rainy season. A photovoltaic module or array can be composed of solar PV cells, which can be organized in series and parallel. The output power of these solar modules is influenced by changing climatic conditions such as irradiance and operating cell temperature. The current work concerns the development of a photovoltaic emulator for testing and conducting research activities on photovoltaic-based systems, avoiding the problems of an intermittent nature dependent on weather conditions [16]. Table 2.2 illustrates the year wise power generation in India from conventional and non-conventional energy sources.

Jansen traveled to Guntur, India in 1868 to observe a solar eclipse. He focused on the lobe of the sun and concluded that it consisted mostly of hydrogen gas that was heated to very high temperatures. However, when he observed the spectrum of the Sun with his spectrograph on 18 August, he found that the wavelengths of the yellow line, indicating the presence of sodium, did not match the wavelengths of sodium. In fact, it does not match the wavelengths of the previously known elements. The line was bright enough; he reasoned that if a way could be found to filter out all but that wavelength of visible light, it should be visible without the aid of an eclipse. Therefore, he invented a spectroscopic helioscope to better analyze the spectrum of the sun [17].

Solar cells, which convert sunlight into electricity, were invented more than a century ago because early solar cells were so inefficient that they were not very useful. Bell Labs researchers developed a primary experimental silicon solar cell in April of 1954.

Figure 2.3 shows the constructional details of a PV cell. A solar PV cell is an ordinary diode, although there are some differences between them. A PV cell has a very thin layer of P-type and thick layer of n-type

Table 2.2 Power generation (MW) from various sources in India annually [15].

Fuel class	In 2013	In 2014	In 2015	In 2016	In 2017	In 2018	In 2019	In 2020	In 2021
Coal	145,273	164,636	185,173	192,163	197,171	200,704	204,224	205,135	209,810
Gas	21,782	23,062	24,509	25,329	24,897	24,937	24,937	249,555	24,924
Diesel	1,199	1,199	993	838	838	638	510	510	510
Nuclear	4,780	5,780	5,780	6,780	6,780	6,780	6,780	6,780	6,780
Hydro	44,335	45,323	47,057	48,858	49,779	49,992	50,047	50,382	51,351
Wind	21,043	23,354	26,777	32,280	34,046	35,626	37,279	37,694	40,083
Solar	2,632	3,744	6,763	12,289	21,651	28,181	32,578	34,628	49,347
Biomass	7,510	7,805	8,110	8,296	8,839	9,242	9,946	10,023	10,610
Total	248,554	274,903	305,162	326,833	344,001	356,100	365,891	370,137	393,415

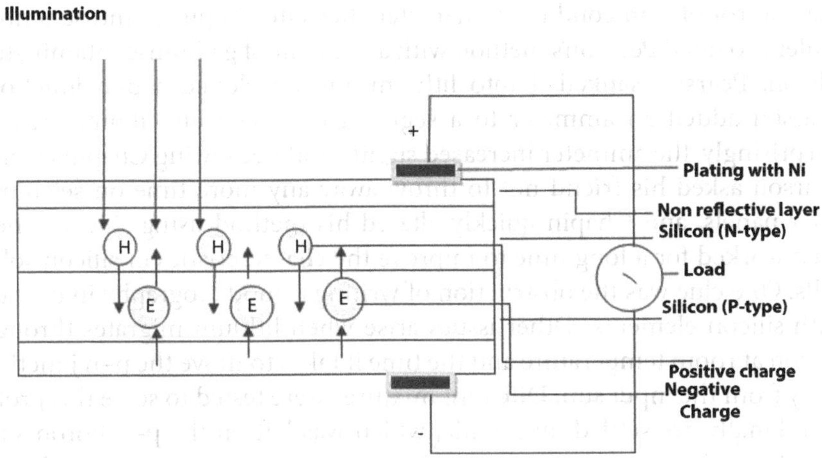

Figure 2.3 Photovoltaic cell.

semiconductor, some additional parts are present in a commercial solar PV cell, which includes non-reflecting material coating and conducting material strips on top of a PV cell.

He worked with metal electrodes in solution and was unable to elaborate on the response. In 1873, Smith developed photocurrent spectroscopy of selenium while using substances for telegraph cables in water. This is impractical because it converts sunlight into electricity. Research on selenium photovoltaics has continued over the following decades. Some applications have been found but have not been installed for a wide range of applications. The next major improvement in solar PV cell technology was made in 1940 by Russell, a semiconductor research scholar at Bell Labs. He examined several silicon samples and identified spots in the center.

On one hand, an extra positive charge is formed. The opposite faces of the potential barrier and the extra negative charge barrier develop an electric field. When the PV solar cell is added to an electrical circuit, photons entering the cell release electrons, which can cool and begin to flow. Ol has received a patent for solar energy. We consider an efficiency of approximately 1%. A group of scientists worked together at Bell Labs 13 years after he built the first practical silicon solar cell. In 1953, Darryl Chapin, working on magnetic materials at Bell's R&D center, attempted to develop a power supply. Chapin studied and selected several alternative energy sources. This is one of the most promising solar energy sources. He experimented with selenium solar panels, but they were inefficient. Currently, the chemist Calvin Fuller and physicist Gerald Pearson are involved in the control of semiconductor characteristics after impurity introduction. Fuller provided Pearson's method with a segment of gallium-contaminated silicon. Pearson sanked it into lithium and developed a p–n junction. Pearson added an ammeter to a segment of silicon and illuminated it. Surprisingly, the ammeter increased significantly. Knowing Chapin's work, Pearson asked his friend not to throw away any more time on selenium solar panels, and Chapin quickly altered his method using silicon. Then then worked for a long time to improve the characteristics of silicon solar cells. One clue was the obstruction of writing a good biography in contact with silicon elements. Other issues arise when lithium migrates through silicon at room temperature and the time it takes to move the p–n junction away from the input sun. Different mixtures were tested to solve this problem. Finally, we settled on arsenic, which was left on the p–n boron surface, boron arsenic was used to make good current contact. Some design improvement is connecting multiple solar cells into what is called a "solar cell." Bell Labs communicated about his development on 25 April 1954, in New Jersey. They introduced the idea of employing solar panels to power a tiny toy Ferris wheel and solar-powered radio station. These silicon solar cells are nearly 6% efficient in the transformation of illumination into electrical energy, which is a significant advancement over all former solar cells. The media wrote that silicon solar PV cells "could mark the beginning of a new era that would ultimately lead to the realization of mankind's most cherished dream of harnessing the sun's nearly limitless energy for civilization purposes." The earliest silicon solar PV cells were costly to manufacture, and the first commercial attempts were not favored. However, within a short period, solar cells have been widely employed in various applications. Chapin quickly simplified the manufacturing process for silicon solar cells and developed a solar cell science experiment for high school students. In 2008, Chapin, Fuller, and Pearson joined the American Inventor Hall of Fame. Solar panels are currently used in all types of devices, from calculators to rooftop solar panels. Modernized designs and recently invented materials have resulted in real-life solar cell production. Reaching efficiencies above 40%, R&D continues to drive cost savings and efficiency gains [18].

Photovoltaic devices are being developed rapidly to achieve electrical energy from the solar photovoltaic method, and this is becoming one of the most efficient alternatives to traditional fossil fuel power generation. However, the first experimental PV devices were shown in the 1950s, and other power generation systems are relatively new compared to technology. In the 1960s, the PV industry made its first major breakthrough in photovoltaic research and development. The industry requires an off-grid system to meet space requirements. However, the cost of PV cells was very high for space requirements than they are today, and it took a decade before the need for off-grid power generation methods was recognized; however, solar cells have replaced the rapidly growing silicon transistors. This resulted in an interesting scientific twist in development. After the oil crisis of the 1970s, the world's attention has focused on the fascination of alternative energy sources for terrestrial use. This has spurred research on photvoltaics as a means of generating terrestrial energy. Although the oil crisis proved short-lived and weakened the economic incentives to develop solar cells, solar cells have emerged as a power generation technology. In 1985, Si PV cells achieved a reasonable efficiency of 20%. In the forthcoming years, the PV cell business had a rising rate of 15%–20%, driven primarily by the international market. In 1997, it recorded an updated, high growth rate. This development has been repeated for decades, and currently, solar cells are not only known as a means of providing electricity and improving the standard for those without access to the grid, but are also used to power the grid. This is also a means to play an important role [19].

2.4 Solar Cell Models

The age of the sun is nothing new. Its history dates to the 7th century BC. Until today, it started by using glass and mirrors to collect the heat of the sun to make a fire. Today, there are sun-powered homes and solar-powered automobiles. Humanity had been getting benefited from solar energy right from the beginning of civilization on Earth, although it was highlighted in 7000 B.C. In fact, solar energy is the basis of all kinds of energy from which we are taking advantage. Solar photovoltaic systems are also a branch of solar energy through which light is converted into electricity. To achieve an efficient solar photovoltaic power system modeling of solar photovoltaic cell is required for studying all possible aspects for building a moderate system. Ideally, a parallel connection of a current source and a diode forms a photovoltaic cell, but it is generally not suitable for a practical photovoltaic cell, as shown in Figure 2.4, which is the most basic ideal photovoltaic solar cell. Among the various models available in the literature, the most frequently employed circuit-based PV model is the single diode model (SDM), which consists of series resistance (Rs), shunt resistance (Rsh), and a linear independent current source in parallel to a diode. In fact, its existence is explained by two resistors in series and parallel, as shown in Figures 2.5 and 2.6 below. This is indicated by the resistance Rsh. These are called Figures 2.5 and 2.6, respectively. The single-diode model is not

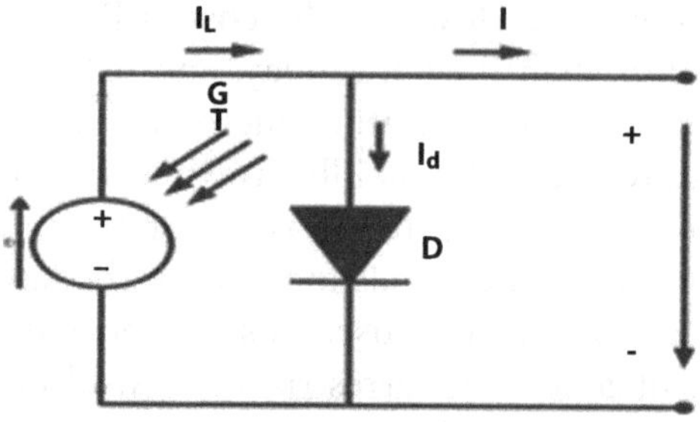

Figure 2.4 Model of a PV cell (ideal) [10].

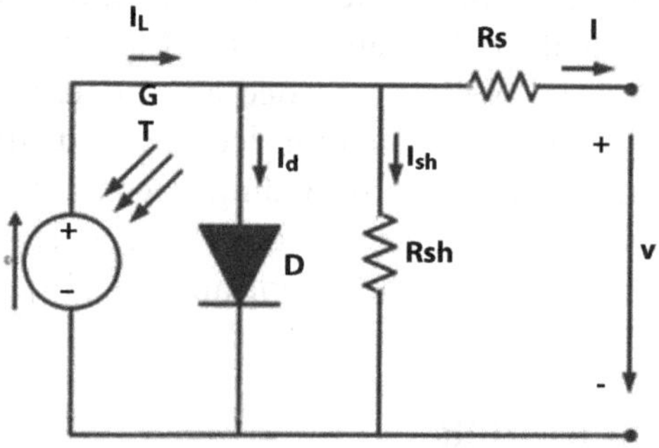

Figure 2.5 Model of PV cell (one diode) [10].

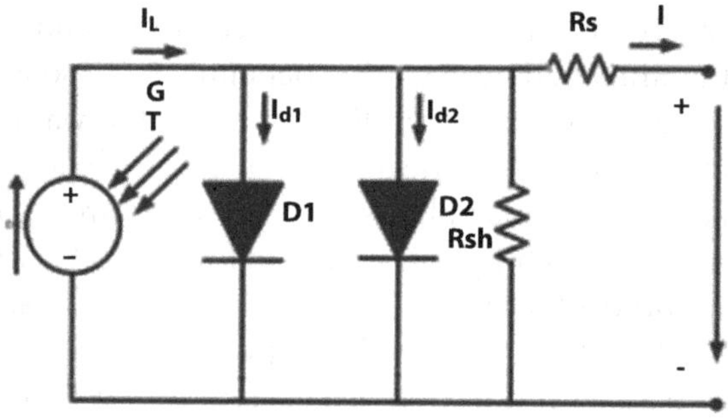

Figure 2.6 Model of solar PV cell (two diode) [10].

complicated to implement, while the two-diode model is more accurate and can more accurately predict the performance of photovoltaic systems. Exploration was performed based on the simulation results, and MATLAB was used for this purpose. In the ideal model, the diode is represented by D, as in the single-diode model, but in the two-diode model, the diodes are represented by D1 and D2 [20].

The equations of the PV model (Figure 2.4) are given by:

$$I = I_{ph} - I_{str}\left[e\frac{V + IR_{ser}}{vth} - 1 \right] - \frac{V + IR_{ser}}{R_{sht}} \tag{2.1}$$

MATLAB was used to simulate the shading conditions of the photovoltaic array. Shade is the impact of clouds, buildings, and trees on the photovoltaic array, which reduces the production of the photovoltaic system. The chromaticity equation can be expressed as:

$$V_{sd} = \left[\frac{(n-1)}{n} \right].V - I(R_{sht} + R_{ser}) \tag{2.2}$$

$$v_{th} = \frac{n_{sc} a_f K_b T_o}{q_e} \tag{2.3}$$

Where v_{th} = voltage of thermal; qe = charge on electron; K_b = constant (Boltzmann); T_o = temperature; n_{sc} = cells connected in series; a_f = diode ideality factor; V = output voltage; I = output current; R_{ser} = series resistance; and R_{sht} = shunt resistance.

2.5 Applications

The year 1923 was the beginning of the history of semiconductor LED for visible light production in the point contact diode of silicon carbide, but it was not much developed, and it became only a part of the solar PV cell history. Industrial-scale production is one of the most important uses of heterostructure solar cells in our country. The technology was shifted to the "Quant" company and from 1974 Ga Al As solar cells were added in most of our satellites and these heterostructure solar PV cells were being used by our space station "MIR" for 15 years [21]. Other applications include laptop computers, telecommunication repeater stations, water pumps, supplying occasional power, cottages and remote residences, parks in remote regions, and navigational aids.

2.6 Conclusion

This chapter details the history of Sharp in developing photovoltaic cells and their application in terrestrial and space communications, where we see that the commercialization of photovoltaic cells has greatly contributed to not only saving energy for society but also to reducing greenhouse gas emissions, as well as providing the community with the benefits of maintenance-free electricity and telecommunications services, making our communities more livable and safer. In the case of advances of PV cell research and development, efficiency increases while prices decrease. As previously mentioned, solar cells with efficiencies of approximately 45% have been developed. It should be noted that prices decrease when stipulation and manufacturing increase. We can therefore conclude that solar cells have the potential to become the single largest energy supply system in the future, much like coal came along in the 19th century industrial revolution. The future of photovoltaic technologies is expected to be bright. During this time, solar energy emerged as a valuable source of unconventional energy, meeting the energy needs of a country's overall development. Harnessing solar energy for universal progress is not a new discovery, but an ancient tradition. As the demand for clean energy increases, there is a need to develop highly efficient photovoltaic (PV) cells. Here, we discuss the early stages of solar energy use, the current boom in photovoltaic development, and the pros and cons of different types of photovoltaic cells.

References

1. https://renewableenergyhub.co.uk/main/solar-panels/the-history-of-solar-power/
2. S. K. Varshney, Z. A. Khan, M. A. Husain and A. Tariq, "A comparative study and investigation of different diode models incorporating the partial shading effects," *2016 International Conference on Electrical, Electronics, and Optimization Techniques (ICEEOT)*, Chennai, India, 2016, pp. 3145-3150, doi: 10.1109/ICEEOT.2016.7755281.
3. Mohammed Aslam Husain, M. Faisal Jalil, M. Tabish S Beg, M. Naseem and Abu Tariq (2012). Modeling and Study of a Standalone PV System Using MATLAB/SIMULINK. *i-manager's J. Electr. Eng.*, 5(4), 30-35. https://doi.org/10.26634/jee.5.4.1863
4. E. Becquerel (1839). "Memory on the electrical effects produced under the influence of solar rays". *Records* 9: 561–567. Issue date: May 7, 1935.
5. W.G. Adams and R.E. Day (1877) "The action of light on selenium" in *Proc. R. Soc.*, A25, 113.

6. Einstein, A, "On the Quantum Theory of Radiation" *Phyikalische Zeitschrift* 18(1917).
7. David C. Brock (Spring 2006). "Useless No More: Gordon K. Teal, Germanium, and Single-Crystal Transistors". *Chemical Heritage Newsmagazine* (ChemicalHeritageFoundation) **24**(1). Retrieved 2008-01-21.
8. Bloch, *F,Z. Phys.* (1928), 52, 555.
9. Wilson, A. H., *Proc. Roy. Soc. A* (1931), 133, 458; 134, 277.
10. P. A. Vladimirovich, "The Main Problem of Photovoltaic Energy and Possible Ways to Solve It: Historical and Technical Approach," *2019 International Conference on Engineering Technologies and Computer Science (EnT)*, 2019, pp. 112-116, doi: 10.1109/EnT.2019.00030.
11. M. A. Green, "Photovoltaics: coming of age," *IEEE Conference on Photovoltaic Specialists*, 1990, pp. 1-8 vol.1, doi: 10.1109/PVSC.1990.111582.
12. M. Symko-Davies, R. L. Mitchell, C. E. Witt, H. P. Thomas, R. King and D. Ruby, "Decade of PV industry R&D advances in silicon module manufacturing," *Conference Record of the Twenty-Eighth IEEE Photovoltaic Specialists Conference - 2000 (Cat. No.00CH37036)*, 2000, pp. 1460-1463, doi: 10.1109/PVSC.2000.916168.
13. P. Mints, "Photovoltaic crystalline and thin film capacity, production, shipments and inventory, 1982–2019; past and future trends," *2015 IEEE 42nd Photovoltaic Specialist Conference (PVSC)*, 2015, pp. 1-3, doi: 10.1109/PVSC.2015.7356251.
14. Fisher, G., Seacrist, M. R., & Standley, R. W. (2012). Silicon Crystal Growth and Wafer Technologies. *Proc. IEEE*, 100(Special Centennial Issue), 1454–1474. doi:10.1109/jproc.2012.2189786
15. S. K. Nag, T. K. Gangopadhyay and J. Paserba, "Solar Photovoltaic: A Brief History of Technologies [History]," in *IEEE Power Energy Mag.*, vol. 20, no. 3, pp. 77-85, May-June 2022, doi: 10.1109/MPE.2022.3150814.
16. Sinha, S., Agarwal, P., Gupta, N. K., Asim, M., & Riyaz, A. (2020). Performance of solar cell under changing atmospheric condition. In *Energy Systems, Drives and Automations: Proceedings of ESDA 2019* (pp. 225-234). Springer Singapore.
17. https://www.aps.org ›apsnews physicshistory
18. https://www.aps.org/publications/apsnews/200904/physicshistory
19. W.G. Adams and R.E. Day (1877) "The action of light on selenium" in *Proc. R. Soc.*, A25, 113.
20. Asim, M., Sarwar, A., Shahabuddin, M., & Manzar, M. S. (2021). Development of solar photovoltaic model for wide range of operating conditions. *Int. J. Power Electron. Drive Syst. (IJPEDS)*, *12*(4), 2483-2491.
21. Z. Alferov, "Heterostructures for Optoelectronics: History and Modern Trends," in *Proc. IEEE*, vol. 101, no. 10, pp. 2176-2182, Oct. 2013, doi: 10.1109/JPROC.2013.2274912.

3

Clustering of Panels and Shade Diffusion Techniques for Partially Shaded PV Array—Review

Vaishali Gautam[1], Shahida Khatoon[1], Mohd Faisal Jalil[2]* and Naimul Hasan[3]

[1]*Department of Electrical Engineering, Jamia Millia Islamia, New Delhi, India*
[2]*Department of Electrical Engineering, Aligarh Muslim University, Aligarh, India*
[3]*Department of Electrical Engineering, Qassim University, Buraydah, Saudi Arabia*

Abstract

Photovoltaic (PV) power generation is primarily dependent on environmental factors. The PV array lifetime and overall system effectiveness are reduced owing to the partial shading condition. Clustering the electrical connections between the solar modules is a viable strategy for minimizing these power losses. This study provides a thorough assessment of different models for clustering and reconfiguring PV arrays in PV systems. These are static and dynamic reconfiguration techniques for extracting the maximum power under the mismatch condition. This chapter explores and analyzes the current breakthroughs in solar PV performance improvement strategies that merit further investigation. Altogether, researchers and academicians working in the field of dedicated solar power generation will benefit from this study. This review aims to encourage additional exploration and advancements in this field with the aim of achieving higher-performing and dependable PV systems.

Keywords: Static reconfiguration, dynamic reconfiguration, PV array, partial shading

**Corresponding author*: mfaisaljalil@gmail.com

Mohammed Aslam Husain, Md Waseem Ahmad, Farhad Ilahi Bakhsh, P. Sanjeevikumar and Hasmat Malik (eds.) Photovoltaic Systems Technology, (43–62) © 2024 Scrivener Publishing LLC

3.1 Introduction

The limited availability of traditional fuels has compelled the world to shift its focus to alternative energy sources. There are numerous options for non-conventional energy sources, including solar photovoltaic, wind, hydro, and tidal [1]. Solar power is considered one of the most promising and abundant sources of renewable energy and is available year-round. It is the purest form of energy, which requires less maintenance than the other forms of energy. However, the effective generation of electricity from solar energy is complex. There is also a problem in handling the inbuilt resistance of PV modules [2]. This problem increases with a non-uniform irradiance level. The non-uniform irradiance level is mainly due to the shadowing of neighboring buildings and trees, movement of clouds, dropping, and dust particles. This phenomenon of non-uniform irradiance is called partial shading. Less current is produced owing to partial shading of the PV cells [3].

Because a single PV module consists of several interconnected cells, when some of the cells are shaded, they still produce current, but they must be equivalent to the current produced by the unshaded cells. This results in the generation of a significant amount of current. The shaded cells act as resistance, causing electricity to be pulled from the unshaded cells. As a result, the temperature within the cells increases, creating hotspots that eventually destroy PV modules [4]. Bypass diodes can be used to resolve hotspots. These bypass diodes, on the other hand, produce many spikes over the PV curve, minimizing power production. For the aforementioned reasons, a gap exists between the installed power production and actual energy generation [5]. Because every plant comprises many converters, implementing these procedures for large-scale power plants is tedious and results in higher costs. To compensate for these drawbacks, the authors devised configuration strategies to boost the power generation. Two major reconfiguration techniques are available: static and dynamic [6].

One way to alleviate shading effects across a solar panel array is through a technique called static reconfiguration, which involves physically rearranging modules without altering their electrical connections. This approach allows the electrical connections between the modules to remain fixed during operation. However, selecting an appropriate layout to distribute the shading impacts evenly throughout the array is the most difficult aspect of this strategy. Unlike dynamic systems, this technology does not require sensors or a switching matrix, thus lowering the cost of the PV system. In a study referenced in [7], the authors created a new connectivity mechanism to distribute shading effects, and this approach was evaluated on a 3 × 3 array, demonstrating superior performance in comparison to "SP, TCT, and BL array" interconnections. Additionally, a Magic Square (MS) puzzle technique was devised to spread partial shading conditions in a 4 × 4 array, as stated in [8]. The magic-square layout improves the maximum power in comparison to "SP, TCT, BL, HC coupled with SP-TCT and BL-TCT array interconnections," according to this study.

Currently, researchers are investigating dynamic reconfiguration techniques that can adjust the connection architecture of PV modules in an array to improve power production when faced with partial shadowing conditions (PSC) and electrical mismatch issues. When the PV modules are shaded, those connected in series have reduced output power and maximum allowable current, which can cause issues. Additionally, a shaded or faulty module may become excessively hot, leading to hotspot phenomena and ultimately module failure [9–14]. Although bypass diodes can prevent hotspots and mismatches, they can result in power losses and localized maxima in the electrical properties of PV modules [15, 16]. Many studies have explored these issues, as noted in references [17–20].

In the study described in [21], the effects of partial shading on a photovoltaic (PV) array were analyzed, and a technique for achieving maximum power point tracking (MPPT) using particle swarm optimization (PSO) was suggested by the authors. The study involved simulating a partially shaded PV array and analyzing the performance of the system using different MPPT techniques. The results demonstrated that the proposed PSO-based MPPT method outperformed the other MPPT methods in terms of efficiency and accuracy in tracing the maximum power point while dealing with different levels of shading. In a study referenced in [22], the authors introduced a unique approach for rearranging solar photovoltaic arrays to improve shade dispersion under partially shaded conditions, using game theory based on magic squares. The authors then compared the performance of their proposed method to other reconfiguration techniques, such as PSO and Genetic Algorithm (GA). The results indicated that the method outperformed PSO and GA significantly in terms of enhancing the shade dispersion factor and increasing output power generation.

This study provides an overview of the current findings on PV reconfiguration alternatives for increasing power output in non-uniform shadowing scenarios. The present study analyzes a variety of PV array reconfiguration techniques for PV systems, including both static and dynamic approaches, to achieve the maximum power output under electrical mismatch conditions. The chapter is structured as follows: Section 3.2 centers on the reconfiguration of PV arrays and covers the PV cell modeling and the explanation of PV reconfiguration. Section 3.3 focuses on the classification of reconfiguration strategies, in which various static and dynamic reconfiguration methods are discussed. Section 3.4 emphasizes the conclusion, followed by references.

3.2 Reconfiguration of PV Array

3.2.1 Modeling of PV Cell

The first and most important stage in studying the characteristics and functionality of any solar PV panel is modeling the PV module. Owing to the non-linear features of PV, accurate solar PV modeling is difficult [23–28]. To conduct a PV array reconfiguration, it is essential to evaluate the electrical properties of the PV module. Numerous models of photovoltaic cells have been investigated over time, with the single-, double-, and triple-diode models being the most widely used [29–31].

Figures 3.1, 3.2, and 3.3 depict the single-, double-, and triple-diode models, respectively, where Irr denotes the current produced by the PV source, Id1 represents the current passing through Diode 1, Id2 is the current passing through Diode 2, and Id3 is the current passing through Diode 3. Rs and Rp are the series and parallel resistance across the diode,

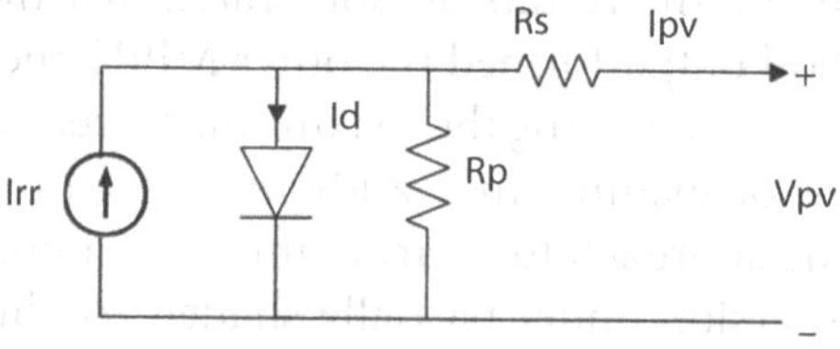

Figure 3.1 Single-diode model.

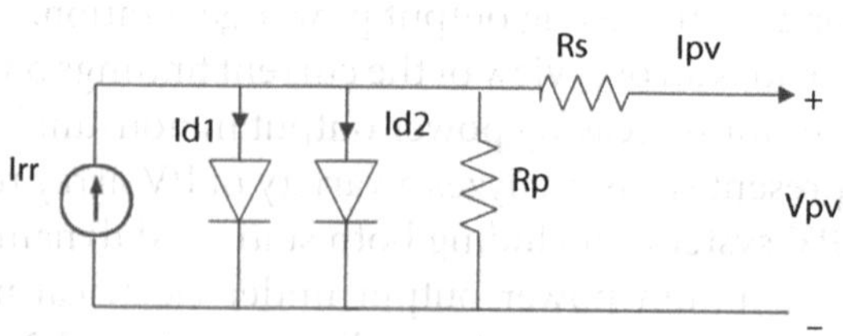

Figure 3.2 Double-diode model.

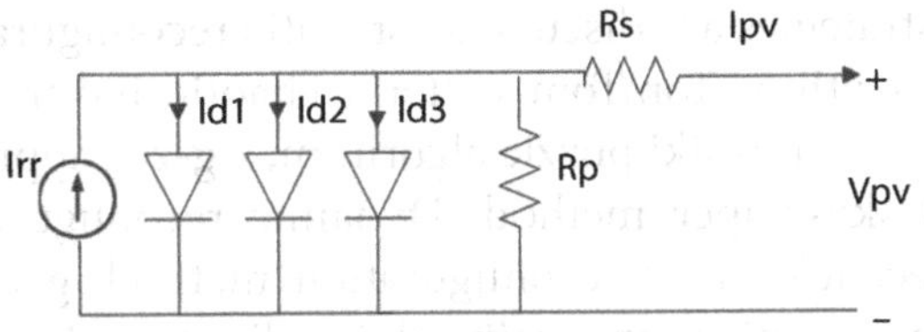

Figure 3.3 Triple-diode model.

respectively. Ipv and Vpv are the current generated in the PV array and voltage generated in the PV array, respectively.

3.2.2 Definition of PV Reconfiguration

The configuration of the PV modules was determined by the arrangement of the PV arrays. To reduce partial-shading losses, many array connectivity topologies have been studied. Reconfigurable PV arrays are another option to compensate for power loss caused by partial shading [32, 33]. Figure 3.4 represents the PV reconfiguration technique.

3.3 Classification of Reconfiguration Strategies

Generally, PV reconfiguration strategies are categorized into two groups: static and dynamic. Various strategies have been proposed in literature [34].

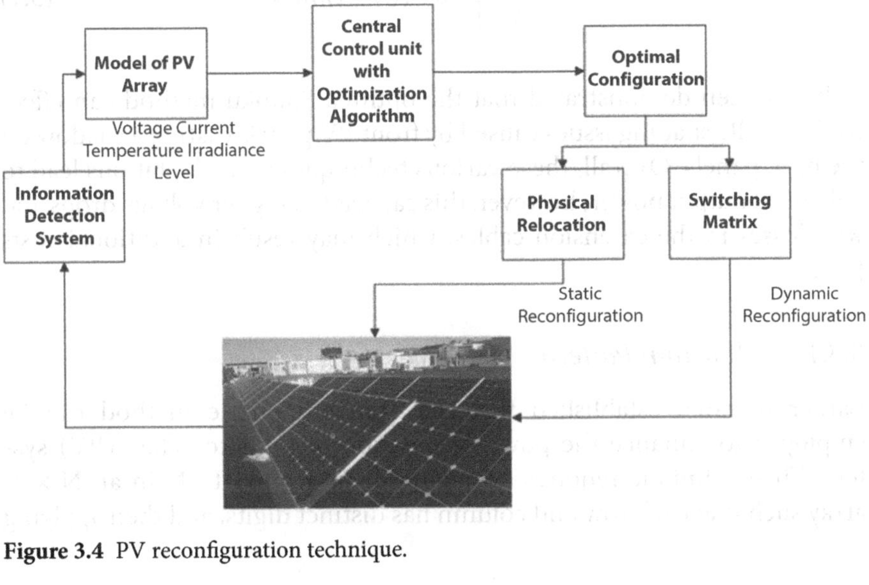

Figure 3.4 PV reconfiguration technique.

Here, various strategies are discussed for static reconfiguration, including the Sudoku algorithm, TomTom pattern, chaotic Baker method, magic square technique, Futoshiki puzzle algorithm, zig-zag approach, odd even approach, and skyscraper method. Dynamic reconfiguration strategies include the electrical array reconfiguration method, genetic algorithm, particle swarm optimization, artificial intelligence algorithm, adaptive array reconfiguration, irradiance reconfiguration by relocation of panels, grasshopper optimization algorithm, and modified Harris hawks optimizer algorithm.

3.3.1 Static Reconfiguration Strategies

3.3.1.1 Sudoku Algorithm

The Sudoku puzzle pattern consists of submatrices that contain digits 1 to 9 without repetition. A reconfiguration method inspired by Sudoku was developed for PV arrays, which enables shaded modules to be distributed across multiple locations [35]. The optimal Sudoku method addresses the limitations of the traditional Sudoku approach, which has poor dispersion. To attain the necessary dispersion factor for PV arrays that have been reconfigured, the magic Sudoku technique includes the positioning of X lattices beneath the PV modules in the 1st column, which will form the nth column, as indicated by Equation (3.1):

$$X = \left\lceil \frac{n-1}{3} \right\rceil + 3 * ((n-1) mod 3) \tag{3.1}$$

It has been demonstrated that the optimal Sudoku method can effectively handle shading issues caused by front PV panels casting a shadow on the back panels. Overall, these various techniques and adaptations lead to enhanced output power; however, this can lead to higher voltage drops and wire losses in the extension cables, which may result in additional costs [36].

3.3.1.2 TomTom Pattern

Earlier research established that the TomTom puzzle method can be employed to enhance the power generation of a photovoltaic (PV) system. This technique requires arranging digits from 1 to N in an N × N array such that each row and column has distinct digits, and then applying specific calculation formulas to generate digits that produce the desired number in the upper left-hand corner of each individual subgrid [37, 38].

3.3.1.3 *Chaotic Baker Method*

The chaotic Baker map (CBM) is a method used in computer engineering to randomly modify the pixel coordinates of a picture for proper communication [39]. The CBM method considers the physical locations of PV panels as pixels and moves them to different positions in the PV array. First, PV arrays of size N × N are partitioned into k rectangles. The CBM technique scans each rectangle after shifting the pixel locations to create a new array row that is then integrated into the new array. One advantage of CBM is that it has an inherently optimal algorithm among a large number of possible variables [40].

3.3.1.4 *Magic Square Technique*

The magic square technique has been designed in previous studies to help in the implementation of the n × n PV array to remedy the evident limitation of the Sudoku algorithm, which can only be conducted in a 9 × 9 array [41]. In any column, row, or diagonal, the summation of the figures is the same in an MS matrix according to a mathematical formula. Moreover, compared with series-parallel (SP), bridge-linked (BL), and TCT, the proposed MS has a lower power loss and a higher fill factor. The MS technique avoids complicated MPPT algorithms, provides better output than the traditional TCT configuration, and produces a smoother PV array characteristic with fewer local maxima, thereby decreasing the mismatch impact [42].

3.3.1.5 *Futoshiki Puzzle Algorithm*

Another form of static reconfiguration method utilizes a puzzle called Futoshiki. This method relies on the idea of physically moving PV panels such that their electrical connections do not change [43]. Each row and column in this puzzle contain numbers 1 to m, without revisiting any of them. The Futoshiki approach allows shading to be dispersed across the overall PV array, resulting in lower power loss. When a mismatch occurs, the PV modules that are the least irradiated disperse, resulting in the minimum current in the row, and the PV array's voltage output remains unchanged for any Futoshiki approach. In comparison to the traditional

TCT configuration under shade conditions, the acquired findings suggest that this technique performs better and reduces the mismatch loss [44].

3.3.1.6 *Zig-Zag Approach*

A different setup for the static reconfiguration method is the novel zig-zag reconfiguration methodology. This procedure entails physically relocating the panels to accommodate a new framework [45]. With the effective application of this method, researchers verified the advantages of the zig-zag method over the TCT structure in terms of power augmentation and shading loss minimization [46].

3.3.1.7 *Odd Even Approach*

To extract the maximum output power, a revised fixed structure approach based on the odd-even method was proposed. Two major factors constitute the odd-even scheme [47]. The initial portion keeps all the electrical connections of the modules intact [48]. In the second part of the strategy, the PV modules are replaced. To achieve the desired increase in the power generated by the PV array, the design was tested with a 4 × 4 array configuration under various mismatch conditions. This method aims to reduce the occurrence of Local Maximum Power Points (LMPPs) for different shading scenarios, making it a better option for MPPT [49].

3.3.1.8 *Skyscraper Method*

To present a different setup for PV array rearrangement, an innovative skyscraper method was developed, replicating the structural design of building groups with varying peaks. This solution depends on a single-time rearrangement of the PV arrays to address the high calculation time and tedious connector links. It also disperses shadowing over the PV array, boosting performance and increasing the power produced [50]. The findings show that more power can be generated by the skyscraper method compared to the TCT and Sudoku methods, while reducing mismatch losses. The main disadvantage of this technology is such that it relies on physical rearrangement, which results in a significant power loss owing to the use of cables. However, it can be performed simply for a maximum array size of 9 × 9 nodes [51].

3.3.2 Dynamic Reconfiguration Strategies

3.3.2.1 Electrical Array Reconfiguration Method

The PV array is redesigned using the EAR approach, which considers the levels of irradiance and row current. In this system, a switching matrix is produced between the PV modules based on the information obtained from a data-collection system that detects the irradiation level [52]. The switching matrix is usually designed using field-programmable gate arrays or digital signal-processor microcontrollers. The EAR method involves the use of a switch matrix between the PV array and centralized inverter, with the inverter serving as a common ground. This allows for the repositioning of PV modules based on information regarding the row current and irradiance level [53].

3.3.2.2 Genetic Algorithm (GA)

The genetic algorithm (GA) is an automatic algorithm that follows Darwin's notion of evolution through selection [54]. The optimization problem for the GA is the generation of power mixed with the weighted current. The trial-and-error method was also applied to acquire the proper parameter limitations. Under wrong conditions, GA, on the other hand, is simple to catch at local optimal solutions. A method called the standard deviation genetic algorithm (SDGA) has been proposed in previous research to address this limitation. The SDGA combines a genetic algorithm with the idea of standard deviation, and employs an objective function calculated by Equations (3.2)–(3.4) to minimize the difference between row currents:

$$\max \text{fitness}(i) = \frac{1}{1+\sigma_1} \tag{3.2}$$

$$\sigma_1 = \sqrt{\frac{1}{N}[\Sigma_{j=1}^{N}(I_j - I_m)^2]} \tag{3.3}$$

$$I_m = \frac{1}{N}\Sigma_{j=1}^{N} I_j \tag{3.4}$$

where the fitness value of the *i*th element of the current population's is fitness (i), the current of each row is represented by I_j, and N denotes the number of rows. In this scenario, the standard deviation of the row currents is equal to 1 [55].

3.3.2.3 Particle Swarm Optimization

PSO is a parallel and randomized search technique that is widely used owing to its versatility and ease of use [56]. In the context of PV array rearrangement to reduce the effects of PSC, PSO is employed with the following goal objective (3.5):

$$\text{maximize } F(i) = \text{sum}(P) + \left(\frac{W_e}{E_e}\right) + (W_p * P_a) \tag{3.5}$$

where F(i) is the *i*th element of the current position's fitness, and the overall power generated by the PV array is the sum (P). The difference between the peak current and the row current is equal to E_e. The power produced by an individual PV panel when it is not bypassed is denoted by P_a. The different weight coefficients are represented as W_e and W_p. PSO decreases local peaks, boosts the output power, and minimizes the row current difference according to comparison trials with the TCT scheme. In addition to the difficulty of adjusting complicated parameters, PSO may cause early closure and mislead system responses [57].

3.3.2.4 Artificial Intelligence Algorithm

Artificial intelligence (AI) techniques for PV array reconfiguration have recently received considerable interest [58]. There are four distinct methods of implementing artificial intelligence algorithms: the feed-forward artificial neural network, fuzzy logic estimator, machine learning algorithm, and clonal selection algorithm. To reconfigure a photovoltaic (PV) array, an approach based on an artificial neural network that utilizes a feed-forward model was proposed. This method utilizes short-circuit currents and irradiance as the primary data for training to identify the most effective configuration. Fuzzy logic is a type of logic that resembles human brain reasoning in making fuzzy, thorough decisions. It is used in systems whose model is uncertain. Machine learning involves the implementation of a cyber-physical system that is capable of advanced signal processing. To develop a technique for generating irradiance profiles, a machine-learning model is utilized for PV array reconfiguration [59].

3.3.2.5 *Adaptive Array Reconfiguration*

Another type of reconfiguration technique is Adaptive Array Reconfiguration, which divides panels into two categories: adaptive and fixed. To ensure irradiance distribution, the current and voltage information of the individual panel obtained by the sensors is communicated to the microcontrol unit [60]. To counteract the negative impacts of shading and create a uniform irradiance distribution across different rows, adaptable panels were rearranged across several rows in accordance with the shadow distribution. However, the AAR approach has some significant drawbacks, including: 1) it is not compatible with low-power PV systems; 2) it requires a sophisticated control mechanism, resulting in an increase in both the system size and cost; and 3) it necessitates a large adaptive portion, as well as additional sensors and switches [61].

3.3.2.6 *Irradiation Equivalence by Relocation of Panels*

By applying the concept of irradiation equivalence, a cluster of photovoltaic (PV) cells is transferred to different rows, ensuring that no current constraints are imposed by any of the series-connected strings or connected PV panels. Instead of isolating shaded panels, they are connected to multiple rows to minimize the mismatch of row currents, similar to how an electric array is reconfigured. The number of solar panels connected in sequence and in parallel remained unaltered during the reconfiguration process. Irradiation is compensated, bypass diode activation is prevented, and thus, the energy production is increased by moving panels in the PV array as per the existing shade circumstances [62]. The computational difficulties of these systems, as well as the need for sensors and switches, are significant drawbacks.

3.3.2.7 *Grasshopper Optimization Algorithm*

The utilization of the grasshopper optimization algorithm is preferred for minimizing the effects of partial shading situations because it is an uncomplicated approach and requires fewer controllable parameters [63]. Different types of shadow patterns like Short Wide (SW), Short Narrow (SN), Large Wide (LW), and Large Narrow (LN) have been examined in existing literature, and their outcomes have been contrasted against TCT, GA, and Sudoku connected PV arrays. Under shade conditions, the GOA increases the power output by 3.361%, 10.949%, 0.864%, and 6.748%, respectively [64].

Table 3.1 Static reconfiguration technique.

S. no.	Techniques	Array size	Complexity	Advantage/ Disadvantage	References
1	Sudoku Algorithm	9 × 9	high	Implementation is easy, sensors are not required	[35, 36]
2	TomTom Pattern	9 × 9	high	Sensors are not required, Tedious rules	[37, 38]
3	Chaotic Baker Method	4 × 4 or 9 × 9	High	No sensors and switches are required, relocation rules are complicated	[39, 40]
4	Magic Square Technique	3 × 3	Low	Minimum power loss, Fixed first column modules	[41, 42]
5	Futoshiki Puzzle Algorithm	5 × 5	Medium	The puzzle has a unique solution and cannot be applied to an array with dimensions other than m=n.	[43, 44]
6	Zig-Zag Approach	4 × 3	Medium	Applicable to every dimension, Flexibility is less	[45, 46]
7	Odd Even Approach	4 × 4	medium	Diminish Local Maximum Power Point, Shade dispersion is not good	[47–49]
8	Skyscraper Method	9 × 9 or 5 × 5	low	Power losses are minimum, Shade dispersion is poor	[50, 51]

Table 3.2 Dynamic reconfiguration technique.

S. no.	Techniques	Array size	Complexity	Advantage/ Disadvantage	References
1	Electrical Array Reconfiguration Method	9 × 9	medium	Methodology is simple, Tedious relocation rules.	[52, 53]
2	Genetic Algorithm	9 × 9	low	Improved power generation, Reliability on weight coefficient is high	[54, 55]
3	Particle Swarm Optimization	9 × 9	low	Better shade dispersion, robust, Local optimization is easy	[56, 57]
4	Artificial Intelligence Algorithm	9 × 9	medium	System is flexible but lot of training data is required	[58, 59]
5	Adaptive Array Reconfiguration	9 × 9	high	Short circuit current is needed only in adaptive part, Computational steps are more	[60, 61]
6	Irradiation Equivalence by Relocation of Panels	9 × 9	medium	Minimum switching requirements, Tedious relocation rules	[62]
7	Grasshopper Optimization Algorithm	9 × 9	medium	Efficient and reliable but require more computing data	[63, 64]
8	Modified Harris hawk optimizer algorithm	9 × 9 or 6 × 4 or 6 × 20	high	Better performance, Tedious search mechanism	[65]

3.3.2.8 Modified Harris Hawk Optimizer Algorithm

In previous studies, a new PV array rearrangement technique based on the modified Harris hawk optimizer (MHHO) was developed as a solution to the limitations of the grasshopper optimization algorithm. This method mimics the predatory strategy of Harris hawks to enhance the functioning of PV arrays under PSCs [65]. A comparison of simulation results between the modified Harris hawk optimizer (MHHO) and other optimization algorithms, including the genetic algorithm and particle swarm optimization, shows that MHHO provides better reconfiguration with significantly fewer iterations than GA and generates more power than other algorithms. Tables 3.1 and 3.2 show a comparison of both the static and dynamic reconfiguration techniques.

3.4 Conclusion

This study provides an overview of many PV array reconfiguration approaches, as well as their variations, which are divided into two groups: static reconfiguration and dynamic reconfiguration. According to the analysis and findings presented, it appears that dynamic reconfiguration methods require the incorporation of more switching equipment, sensors, and advanced control algorithms to adjust the connections between solar panels in a photovoltaic (PV) array. However, the static reconfiguration technique does not require supplementary switches or sensors for a uniform distribution of shading across the PV array. However, because these methods have a low degree of shaded dispersion, identifying the optimal PV array reconfiguration under shadowing conditions is challenging. This chapter highlights the information and defined characteristics, and discusses the benefits and drawbacks of various reconfiguration strategies. In summary, this literature review emphasizes the significance of utilizing clustering and shade-diffusion methods to enhance the efficiency of partially shaded photovoltaic (PV) arrays. It also sheds light on the recent progress made in this domain. We hope that this review will stimulate further research and development in this area and contribute to the realization of more efficient and reliable PV systems.

References

1. S. Husain and N. A. Shrivastava, "A Comparative Analysis of Multi-objective Optimization Algorithms for Stand-Alone Hybrid Renewable Energy

System," *2020 2nd International Conference on Innovative Mechanisms for Industry Applications (ICIMIA)*, 2020, pp. 255-260.
2. T. S. Babu, N. Rajasekar, K. Sangeetha, "Modified particle swarm optimization technique based maximum power point tracking for uniform and under partial shading condition," *Appl. Soft Comput.* 34 (2015) 613–624.
3. M. F. Jalil, S. Khatoon, I. Nasiruddin, and R. C. Bansal, "An Improved Feasibility Analysis of Photovoltaic Array Configurations and Reconfiguration Under Partial Shading Conditions," *Electr. Power Components Syst.*, vol. 48, no. 9–10, pp. 1077–1089, 2020.
4. A. S. Yadav, V. Mukherjee, "Line losses reduction techniques in puzzled PV array configuration under different shading conditions," *Sol. Energy*, Volume 171, 2018, Pages 774-783, ISSN 0038-092X.
5. G. Meerimatha, B. L. Rao, "Novel reconfiguration approach to reduce line losses of the photovoltaic array under various shading conditions," *Energy*, Volume 196, 2020, 117120, ISSN 0360-5442.
6. B. Dhanalakshmi, N. Rajasekar, "A novel Competence Square based PV array reconfiguration technique for solar PV maximum power extraction," *Energy Convers. Manage.*, Volume 174, 2018, Pages 897-912, ISSN 0196-8904.
7. P. Srinivasa Rao, G. Saravana Ilango and C. Nagamani, "Maximum Power from PV Arrays Using a Fixed Configuration Under Different Shading Conditions," in *IEEE J. Photovoltaics*, vol. 4, no. 2, pp. 679-686, March 2014.
8. A. S. Yadav, R. K. Pachauri, Y. K. Chauhan, S. Choudhury, R. Singh, "Performance enhancement of partially shaded PV array using novel shade dispersion effect on magic-square puzzle configuration," *Sol. Energy*, Volume 144, 2017, Pages 780-797, ISSN 0038-092X.
9. C. Deline, A. Dobos, S. Janzou, J. Meydbray, M. Donovan, "A simplified model of uniform shading in large photovoltaic arrays," *Sol. Energy*, Volume 96, 2013, Pages 274-282, ISSN 0038-092X.
10. M. C. A. García, W. Herrmann,W. Böhmer B. Proisy, "Thermal and electrical effects caused by outdoor hot-spot testing in associations of photovoltaic cells," *Prog Photovolt: Res Appl*, 2003;11(5):293–307.
11. D. Giaffreda, M. Omaña, D. Rossi and C. Metra, "Model for Thermal Behavior of Shaded Photovoltaic Cells under Hot-Spot Condition," *2011 IEEE International Symposium on Defect and Fault Tolerance in VLSI and Nanotechnology Systems*, 2011, pp. 252-258.
12. G. Schirripa Spagnolo, P. Del Vecchio, G. Makary, D. Papalillo and A. Martocchia, "A review of IR thermography applied to PV systems," *2012 11th International Conference on Environment and Electrical Engineering*, 2012, pp. 879-884.
13. W. Herrmann, W. Wiesner and W. Vaassen, "Hot spot investigations on PV modules-new concepts for a test standard and consequences for module design with respect to bypass diodes," *Conference Record of the Twenty Sixth IEEE Photovoltaic Specialists Conference - 1997*, 1997, pp. 1129-1132

14. E. L. Meyer and E. E. van Dyk, "Assessing the reliability and degradation of photovoltaic module performance parameters," in *IEEE Trans. Reliab.*, vol. 53, no. 1, pp. 83-92, March 2004.
15. R. Candela, V. di Dio, E. R. Sanseverino and P. Romano, "Reconfiguration Techniques of Partial Shaded PV Systems for the Maximization of Electrical Energy Production," *2007 International Conference on Clean Electrical Power*, 2007, pp. 716-719.
16. S. Silvestre, A. Boronat, A. Chouder, Study of bypass diodes configuration on PV modules, *Appl. Energy*, Volume 86, Issue 9, 2009, Pages 1632-1640, ISSN 0306-2619
17. N. Gokmen, E. Karatepe, F. Ugranli, S. Silvestre, Voltage band based global MPPT controller for photovoltaic systems, *Sol. Energy*, Volume 98, Part C, 2013, Pages 322-334, ISSN 0038-092X.
18. M. Naseem, M. A. Husain, A. F. Minai, A. N. Khan, M. Amir, J. Dinesh Kumar, and A. Iqbal, "Assessment of meta-heuristic and classical methods for GMPPT of PV system," *Trans. Electr. Electron. Mater.*, 22, 217-234, 2021.
19. M. A. Husain, A. Jain, A. Tariq, and A. Iqbal, "Fast and precise global maximum power point tracking techniques for photovoltaic system," *IET Renewable Power Gener.*, 13(14), 2569-2579, 2019.
20. M. Naseem, M. A. Husain, J. D. Kumar, A. F. Minai, A. Ahmad, S. M. Ali, and A. S. Ansari, "A Spider Monkey Optimization Based Global Maximum Power Point Tracking Technique for Photovoltaic Systems," In *2022 2nd International Conference on Emerging Frontiers in Electrical and Electronic Technologies (ICEFEET)* (pp. 1-6), IEEE, 2022.
21. M. Nisha, M. Andleeb, F. I. Bakhsh, "Effect of Partial Shading on a PV Array and Its Maximum Power Point Tracking Using Particle Swarm Optimization," *J. Phys. Conf. Ser.*, vol. 1817, no. 1, pp. 1-14, March 2021.
22. R. K. Pachauri, R. Swain, A. F. Minai, F. I. Bakhsh, "Improved Magic Square Game theory based SPV Array Reconfiguration Method to Improve Shade Dispersion Factor under PSCs", *2022 1st International Conference on Sustainable Technology for Power and Energy Systems*, NIT Srinagar, India, July 4-8, 2022.
23. M. Azharuddin, T. S. Babu, N. Bilakanti, N. Rajasekar, "A nearly accurate solar photovoltaic emulator using a dspace controller for real-time control", *Electric Power Components and Systems*, 44 (2016) 774–782.
24. N. Rajasekar, N. K. Kumar, R. Venugopalan, "Bacterial foraging algorithm based solar pv parameter estimation," *Sol. Energy*, 97 (2013) 255–265.
25. T. S. Babu, J. P. Ram, K. Sangeetha, A. Laudani, N. Rajasekar, "Parameter extraction of two diode solar pv model using fireworks algorithm," *Sol. Energy*, 140 (2016) 265–276.
26. D. Alam, D. Yousri, M. Eteiba, "Flower pollination algorithm based solar pv parameter estimation," *Energy Convers. Manage.*, 101 (2015) 410–422.

27. B. Jacob, K. Balasubramanian, S. M. Azharuddin, N. Rajasekar, *et al.*, "Solar pv modelling and parameter extraction using artificial immune system," *Energy Procedia*, 75 (2015) 331–336.
28. B. Jacob, K. Balasubramanian, T. S. Babu, N. Rajasekar, "Parameter extraction of solar pv double diode model using artificial immune system," in: *2015 IEEE international conference on signal processing, informatics, communication and energy systems (SPICES)*, IEEE, pp. 1–5.
29. D. Yousri, T. S. Babu, D. Allam, V. K. Ramachandaramurthy, M. Eteiba, "Fractional chaotic ensemble particle swarm optimizer for identifying the single, double, and three diode photovoltaic models' parameters," *Energy*, 2020, 116979.
30. D. Allam, D. Yousri, M. Eteiba, "Parameters extraction of the three diode model for the multicrystalline solar cell/module using moth-flame optimization algorithm," *Energy Convers. Manage.*, 123, 2016, 535–548.
31. V. Khanna, B. Das, D. Bisht, P. Singh, *et al.*, "A three diode model for industrial solar cells and estimation of solar cell parameters using pso algorithm," *Renewable Energy*, 78, 2015, 105–113.
32. S. Malathy, R. Ramaprabha, "Reconfiguration strategies to extract maximum power from photovoltaic array under partially shaded conditions," *Renewable Sustainable Energy Rev.*, Volume 81, Part 2, 2018, Pages 2922-2934, ISSN 1364-0321.
33. La Manna D, Li Vigni V, Riva Sanseverino E, Di Dio V, Romano P., "Reconfigurable electrical interconnection strategies for photovoltaic arrays: a review," *Renew Sustain Energy Rev*, 2014,33:412–26.
34. F. Belhachat, C. Larbes, "PV array reconfiguration techniques for maximum power optimization under partial shading conditions: A review," *Sol. Energy*, Volume 230, 2021, Pages 558-582, ISSN 0038-092X.
35. S. Anjum, V. Mukherjee, and G. Mehta, "Advanced SuDoKu-Based Reconfiguration Strategies for Maximum Power Extraction From Partially Shaded Solar Photovoltaic Array," ASME. *J. Sol. Energy Eng.*, Dec 2021; 143(6): 061003.
36. K. Rajani and T. Ramesh, "Maximum power enhancement under partial shadings using a modified Sudoku reconfiguration," in *CSEE J. Power Energy Syst*, vol. 7, no. 6, pp. 1187-1201, Nov. 2021.
37. Tatabhatla VMR, Agarwal A, Kanumuri T., "Minimising the power loss of solar photo voltaic array through efficient reconfiguration of panels," *Proc. Inst. Mech. Eng., Part A: J. Power Energy*, 2019; 234(5):690–708.
38. V. M. R. Tatabhatla, A. Agarwal and T. Kanumuri, "Improving the output power of Solar PV Array under different irradiation conditions," *2018 IEEE International Students' Conference on Electrical, Electronics and Computer Science (SCEECS)*, 2018, pp. 1-6.
39. V.M.R. Tatabhatla, A. Agarwal, & T. Kanumuri, "A generalized chaotic baker map configuration for reducing the power loss under shading conditions," *Electr Eng*, **102,** 2227–2244, 2020.

40. V. M. R. Tatabhatla, A. Agarwal and T. Kanumuri, "A Chaos Map Based Reconfiguration of Solar Array to Mitigate the Effects of Partial Shading," in *IEEE Trans. Energy Convers.*, vol. 37, no. 2, pp. 811-823, June 2022.
41. G. H. K. Varma, V. R. Barry & R. K. Jain, "A Novel Magic Square Based Physical Reconfiguration for Power Enhancement in Larger Size Photovoltaic Array," *IETE J. Res.*, 2021.
42. M. Etarhouni, B.Chong, L. Zhang, "A novel square algorithm for maximising the output power from a partially shaded photovoltaic array system," *Optik*, Volume 257, 2022, 168870, ISSN 0030-4026.
43. K. A. Bapurao, S. Mikkili & P. K. Bonthagorla, "A Review on Static Reconfiguration Techniques of Solar PV to Mitigate Mismatch Loss and Minimize Partial Shading Effect," *IETE J. Res.*, 2020.
44. G. S. Krishna, T. Moger, "Reconfiguration strategies for reducing partial shading effects in photovoltaic arrays: State of the art." *Sol. Energy*, Volume 182, 2019, Pages 429-452, ISSN 0038-092X.
45. Vijayalekshmy S, Bindu GR, Rama IS, "A novel zig-zag scheme for power enhancement of partially shaded solar arrays," *Sol Energy*, 2016, 135:92–102.
46. S. Vijayalekshmy, G. R. Bindu and S. R. Iyer, "Performance comparison of Zig-Zag and Su do Ku schemes in a partially shaded photo voltaic array under static shadow conditions," *2017 Innovations in Power and Advanced Computing Technologies (i-PACT)*, 2017, pp. 1-6.
47. I. Nasiruddin, S. Khatoon, M. F. Jalil, R.C. Bansal, "Shade diffusion of partial shaded PV array by using odd-even structure," *Sol. Energy*, Volume 181, 2019, Pages 519-529, ISSN 0038-092X.
48. I. Nasiruddin, S. Khatoon, M. F. Jalil, and R. C. Bansal, "Shade diffusion of partial shaded PV array by using odd-even structure," *Sol. Energy*, vol. 181, pp. 519–529, Mar. 2019.
49. K. Yadav, B. Kumar, D. Swaroop, "Mitigation of Mismatch Power Losses of PV Array under Partial Shading Condition using novel Odd Even Configuration," *Energy Rep.*, Volume 6, 2020, Pages 427-437, ISSN 2352-4847.
50. M. S. S. Nihanth, J. P. Ram, D. S. Pillai, A. M.Y.M. Ghias, A. Garg, N. Rajasekar, "Enhanced power production in PV arrays using a new skyscraper puzzle based one-time reconfiguration procedure under partial shade conditions (PSCs)," *Sol. Energy*, Volume 194, 2019, Pages 209-224, ISSN 0038-092X.
51. G.S. Krishna, T. Moger, "Enhancement of maximum power output through reconfiguration techniques under non-uniform irradiance conditions," *Energy*, Volume 187, 2019, 115917, ISSN 0360-5442.
52. G. Velasco-Quesada, F. Guinjoan-Gispert, R. Pique-Lopez, M. Roman-Lumbreras and A. Conesa-Roca, "Electrical PV Array Reconfiguration Strategy for Energy Extraction Improvement in Grid-Connected PV Systems," in *IEEE Trans. Ind. Electron.*, vol. 56, no. 11, pp. 4319-4331, Nov. 2009.
53. A. M. Ajmal, V. K. Ramachandaramurthy, A. Naderipour, J.B. Ekanayake, "Comparative analysis of two-step GA-based PV array reconfiguration

technique and other reconfiguration techniques," *Energy Convers. Manage.*, Volume 230, 2021, 113806, ISSN 0196-8904.
54. M. D. Flórez, D. G. Montoya, L. A. T. Grisales, C.A.R. Paja, "PV Array Reconfiguration Based on Genetic Algorithm for Maximum Power Extraction and Energy Impact Analysis," *Sustainability,* **2022**, *14*, 3764.
55. N. A. Rajan, K. D. Shrikant, B. Dhanalakshmi, N. Rajasekar, "Solar PV array reconfiguration using the concept of Standard deviation and Genetic Algorithm," *Energy Procedia*, Volume 117, 2017, Pages 1062-1069, ISSN 1876-6102.
56. T. S. Babu, J. P. Ram, T. Dragičević, M. Miyatake, F. Blaabjerg and N. Rajasekar, "Particle Swarm Optimization Based Solar PV Array Reconfiguration of the Maximum Power Extraction Under Partial Shading Conditions," in *IEEE Trans. Sustainable Energy*, vol. 9, no. 1, pp. 74-85, Jan. 2018.
57. P. T. Sawant and C. L. Bhattar, "Optimization of PV System Using Particle Swarm Algorithm under Dynamic Weather Conditions," *2016 IEEE 6th International Conference on Advanced Computing (IACC)*, 2016, pp. 208-213.
58. A. Youssef, M. E. Telbany, A. Zekry, "Reconfigurable generic FPGA implementation of fuzzy logic controller for MPPT of PV systems," *Renewable Sustainable Energy Rev.*, Volume 82, Part 1, 2018, Pages 1313-1319, ISSN 1364-0321.
59. M. Mosbah, R. Zine, M. Hatti, S.H. Oudjana, S. Arif, "Network Reconfiguration Management in Intelligent Distribution System Taking into Account PV Production Variation Using Grey Wolf Optimizer," In: Hatti, M. (eds) *Artificial Intelligence and Renewables Towards an Energy Transition*, ICAIRES 2020, Lecture Notes in Networks and Systems, vol 174, Springer, Cham.
60. G. S. Krishna, T. Moger, "A novel adaptive dynamic photovoltaic reconfiguration system to mitigate mismatch effects," *Renewable Sustainable Energy Rev.*, Volume 141, 2021, 110754, ISSN 1364-0321.
61. S. Saravanan, R. S. Kumar, A. Prakash, T. Chinnadurai, R. Tiwari, N. Prabaharan, B. C. Babu, "Chapter 8 -Photovoltaic array reconfiguration to extract maximum power under partially shaded conditions, *Distributed Energy Resources in Microgrids*, Academic Press, 2019, Pages 215-241, ISBN 9780128177747.
62. G. Velasco-Quesada, F. Guinjoan-Gispert, R. Piqué-López, M. Román-Lumbreras, A. Conesa-Roca, "Electrical pv array reconfiguration strategy for energy extraction improvement in grid connected pv systems," *IEEE Trans. Ind. Electron.*, 56, 2009, 4319–4331.
63. A. A. Ghavifekr, A. Mohammadzadeh and J. F. Ardashir, "Optimal Placement and Sizing of Energy-related Devices in Microgrids Using Grasshopper Optimization Algorithm," *2021 12th Power Electronics, Drive Systems, and Technologies Conference (PEDSTC)*, 2021, pp. 1-4.

64. R. Sridhar, C. Subramani, S. Pathy, "A grasshopper optimization algorithm aided maximum power point tracking for partially shaded photovoltaic systems," *Comput. Electr. Eng.*, Volume 92, 2021, 107124, ISSN 0045-7906.
65. D. Yousri, D. Allam, M. B. Eteiba, "Optimal photovoltaic array reconfiguration for alleviating the partial shading influence based on a modified harris hawks optimizer," *Energy Convers. Manage.*, Volume 206, 2020, 112470, ISSN 0196-8904.

4

Advances in Solar PV-Powered Electric Vehicle Charging System

Rashid Ahmed Khan[1*], Noorul Islam[2], Seerin Ahmad[3], Bushra Sabir[1], Mohammed Aslam Husain[4] and Hwa-Dong Liu[1,5]

[1]*Department of Electrical Engineering, National Taiwan University of Science and Technology, Taipei, Taiwan*
[2]*Department of Electrical Engineering, Z.H. College of Engg. & Technology, Aligarh Muslim University, Aligarh, India*
[3]*Department of Electrical Engineering and Computer Science, Texas A&M University, Kingsville, USA*
[4]*Electrical Engineering Department, Rajkiya Engineering College Ambedkar Nagar, Akbarpur, India*
[5]*Undergraduate Program of Vehicle and Energy Engineering, National Taiwan Normal University, Taipei, Taiwan*

Abstract
The alarming global warming situation has led to the adoption of transportation infrastructure based on renewable energy. Electric vehicles (EVs) are a green form of transportation that eradicate greenhouse gas emissions. A PV system-based electric vehicle charging system is a viable step towards sustainability because solar energy has great potential for deriving power from PV panels. This report presents a comprehensive study on the worldwide deployment of solar PV-EV systems and modern solar-assisted electric vehicle charging systems. Additionally, this article discusses various topics related to electric vehicles, including their evolution of electric vehicles, various types of charging stations, and various innovative energy storage methods. Different hybrid storage systems have also been discussed to reduce the size of the battery, as well as various battery life issues and solutions to some of these issues. This study also covers how current advancements in battery storage size can be achieved, as well as the efficiency, sustainability, and cost. However, the electrifying trend in the automobile sector has recently transformed the industry. Electric vehicles have a promising future if appropriate legislation

**Corresponding author*: rashidkhan6417@gmail.com

Mohammed Aslam Husain, Md Waseem Ahmad, Farhad Ilahi Bakhsh, P. Sanjeevikumar and Hasmat Malik (eds.) Photovoltaic Systems Technology, (63–84) and government help are enacted, as well as technological breakthroughs in the field.

Keywords: Battery energy storage system (BESS), battery management system (BMS), electric vehicle (EV), GHG (Green House Gases), maximum power point tracker (MPPT), photovoltaic (PV) module, renewable energy resources (RES)

4.1 Introduction

The automotive sector has given importance to the lifespan of communities and plays a significant role in most countries. Traditional automobiles are usually disastrous to the environment because they require a large amount of fossil fuels to develop kinetic energy [1]. Concerns about environmental change and the security of the petro energy supply are paving the way for renewable energy systems (RESs) to replace traditional transportation energy types. To maintain environmental sustainability, more EV innovations are being developed [2]. EVs are the most practical and environmentally friendly mode of transportation, and in many cities worldwide, they are becoming more interwoven [3, 4]. In several countries, having a certain number of electric cars (EVs) is legally obligatory. The utilization of General Electric equipment and power consumption are linked. By 2022, the number of fossil fuel-powered vehicles will have surpassed 35 million, increasing greenhouse gas emissions, particularly in countries lacking natural alternatives [5], while annual EV sales in the USA (United States of America) are expected to reach 2.4 million by 2024 [6]. However, in many nations EV technologies lack an appropriate and comprehensive recycling structure. Energy consumption is linked to the use of General Electric equipment and electric vehicles (EVs), which leads to a surge in GHG emissions, particularly in countries that lack alternate natural supplies [7]. The operating cost is still a concern, and the battery charging technique has not progressed sufficiently to be durable. This raises a slew of adjustments in important development sectors such as networking, emissions, and enterprise launching on a broad scale. Some of the challenges faced by the industry are establishing charge–discharge time, EV autonomy, battery degradation, vehicle speed, and battery longevity. The usage of rechargeable batteries in electric vehicles has become increasingly common in recent years [8, 9]. EVs have been tested and used in several energy storage technologies, including NiMH, lithium-ion, and lead-acid batteries [10]. For example, lithium-ion batteries have received extensive acceptance owing to their excellent efficiency, high energy density, and extended lifetime [11, 12]. Due to the lucrative operation of this battery innovation, numerous attempts to strengthen the resilience and reliability of lithium-ion batteries have already been made [13–15]. From a full-cycle perspective, charging automobiles with renewable energy sources is better for attaining clean mobility with zero emissions and environmental impact. Furthermore, when these EVs are charged with electrical power derived from fossil fuel-based sources, they may emit the same amount of GHG and have the same or even greater environmental impact than regular traditional gasoline vehicles [16–19]. Using RESs to charge EVs can help reduce GHG emissions while lowering the charging prices. In contrast, wind and photovoltaic energy change dramatically over time, owing to their greater reliance on weather conditions. Intermittency and nondispatchable qualities are common characteristics [20–24]. These characteristics render coordinating EV charging with renewable generation and grid loads extremely difficult. The potential consequences of the EV charging demand on power systems have been studied extensively.

This chapter is divided into into nine sections, and the introduction is addressed in Section 4.1. A brief overview of the electric vehicle charging system is presented in Section 4.2. The evolution of EVs is presented in Section 4.3. Section 4.4 gives the classification of the different EV charging stations. Different approaches for charging EVs are explained in Section 4.5. The recharging and innovative methods are discussed in Section 4.6. In Section 4.7, energy storage for EV charging is explained. Hybrid energy storage technologies are presented in Section 4.8. In Section 4.9, the battery management system is discussed, and the conclusion and future aspects are presented at the end of the chapter.

4.2 Overview of Electric Vehicle (EV) Charging System

The popularity of electric vehicles (EVs) is increasing rapidly worldwide. Electric automobiles are ecologically favorable and energy efficient compared to internal combustible engine vehicles (ICEVs). The electric vehicle cannot be widely adopted owing to the deficiency of charging stations for EVs. Because of the increase in the adoption of electric vehicles (EVs), more charging stations are being installed in public places [25–27]. If fossil fuel powers the existing utility grid and is used for charging the EV, it would adversely affect the existing distribution system and it would not be environmentally friendly. As the potential for generating electrical energy from the solar PV panel is greater than that of solar insolation, this generation can be integrated with electric vehicle charging. This is beneficial for an eco-friendly and sustainable environment.

Accepting the significance of the electric vehicle, worldwide international and national government agencies of the United Kingdom (UK), China, the United States of America (USA), France, Norway, Germany, and many more have passed different commitments and administrative steps and allowed various funds to encourage the implementation of electric vehicle (EV) [28]. To avoid the global climate change scenario, it is expected by the long-term planning scenario that by the year 2050, EVs will replace almost the entire vehicle transport, which would be powered by green resources [29].

The advantages of promoting Electric Vehicle (EV) solar PV charging systems are as follows:

i. Trimming the EVs charging load on the utility.
ii. The voltage issue in the distribution system was rectified.
iii. The charges of the electrical supply remunerated to utilities are minimized.
iv. In direct dc EV-PV, the interconnection charging system has a higher efficiency.
v. The EV battery doubles the Solar PV storage, so the storage of energy is minimized.
vi. Suitability of V2G (vehicle to the grid) and V2H (Vehicle to the home) approaches.
vii. Reduction in the cost of fuel and no emissions at the tail pipe.
viii. The absence of moving parts resulted in no maintenance or noise.
ix. Feasible to implement anywhere.

The development of battery technologies is a key point for the electric vehicle acceptance worldwide, and to meet the power requirements of these EVs, there is a need for facilities. To promote EVs, it is necessary to establish charging infrastructure stations and advanced equipment.

4.3 Evolution of Electric Vehicles

Electric vehicles (EVs) can be defined as vehicles that are electrically driven and employ one or more electric motors that help in their propagation.

EVs may include electric trains, cars, buses, scooters, and motorbikes. In this definition, EVs are restricted to plug-in hybrid electric vehicle (PHEV), purely battery electric vehicle (BEV), and hybrid electric vehicle (HEV). Very little is known about the operation of electric vehicles before that of Internal Combustion Engine Vehicles (ICEV) [30]. In 1828, the first EV was commissioned, and in the late 19th century, it experienced development at its peak. At the beginning of the 20th century, the advancement in Internal Combustion Engine Vehicles (ICEV) with good refilling infrastructure, low-cost gasoline, short refueling times, and large driving ranges led to a deterioration in the production of electric vehicles. Overall, it can be concluded that EVs have several merits such as no smell or vibration, and there is an ease in the manipulation of the gear as compared to gasoline vehicles. The fundamental cause of the revival of electric vehicles in the latter half of the twentieth century was the environmental impact of internal combustion engine vehicles (ICEVs) and the rapid rise in oil prices. Electric vehicles can cause serious inroads to the vehicle industry owing to the latest developments in power electronics, controllers, converters, battery technologies, and microelectronics.

4.4 Classification of Electric Vehicle (EV) Charging Stations

Environmental concerns caused by the emission of greenhouse gases by internal combustion engine vehicles (ICEVs) is a primary element driving the growth of the electric vehicle sector. Additionally, the prospects of EVs have inaugurated the integration of electric vehicle charging systems with solar PV panels. The bridging factor between the two segments is battery recharging. Recharging stations were briefly categorized into four basic sectors.

4.4.1 Residential/Home Charging Station

During the overnight tenure, when the electric vehicle users return to their home, they plug in their cars. The residential charging station does not require any metering or user authentication, and it may also require a customized circuit. There may be a portable charging station on the wall.

4.4.2 Public Charging Station

A public charging station is a charging station that may be present at a rest stop to permit EV users to charge their vehicles during a long-term journey.

A straphanger in the metropolitan city may also avail of this service regularly, and the user may also it to charge their cars while parking for a longer or shorter tenure.

4.4.3 Charging During Park

Usually, in a commercial enterprise where the owner of the parking lot offers partnership, these charging stations may have a higher or slower charging speed. These stations encourage EV users to charge their vehicles while working at nearby facilities. parking at the mall and centers, and business avenues also fall in this category.

4.4.4 Fifteen Minutes Less Charging or Charging Swabs

Battery swabs of automobiles energized by hydrogen fuel cells and electricity can be used to implement this charging system. It considers tallying the refueling expectations of regular users. This is feasible because the potential to manage fast-charging and battery capacity is increasing. The solar PV module roofs the charging station, and the output electrical power required to charge the electric vehicle can be obtained primarily during daytime business hours. The automobile was sitting in a parking lot without reason under the subjection of solar insolation. The charging of electric vehicles (EVs) by solar PV panels is further enhanced by advancing the battery management system and power conversion technology and improving installation practices by applying appropriate design standards [31, 32].

4.5 Approaches to PV-EV Charging System

In this section, we briefly discuss two charging approaches:

i. Solar PV grid charging system
ii. Solar PV standalone charging system.

The PV grid recharging strategy has the merit that when there is insufficient solar insolation, the electric vehicle is charged through the grid (utility) supply system. It has one more advantage when there is no electric vehicle at the charging station and the solar PV supply may be injected into the utility (grid). While considering the solar PV-standalone approach, it has advantages in remote areas such as deserts, islands, and areas that do not receive the supply from the grid (utility) or where the cost of grid (utility) supply is greater. The layout of the PV Standalone charging system is simpler because the requirement for power conversion stages is lower [33].

4.5.1 Solar PV Grid-Charging Station

The incorporation of electric vehicle (EV) charging increases the burden on electrical utilities or grids. This is because for the fast charging of EVs, a large amount of current is extracted from the utilities (grids). Moreover, as per the premium tariff rate, owners must pay hefty charges during busy hours if the user charges the EV. Hence, this charging tariff rate is applied to reduce the burden on the grids, because these grids are the principal dynamic segments of the utility. To minimize the burden on the electrical utility (grid), it is necessary to implement solar PV grid-charging stations [34]. Because of its varied features, the solar PV grid charging station is likely to attract more electric vehicle users. A solar PV grid charging system (except the solar PV module and electric grid) includes three major parts, which are listed below:

I. DC–DC MPPT converter (a combination of DC–DC power converter along with maximum power point tracker (MPPT))
II. DC charger (bidirectional)
III. Inverter (bidirectional).

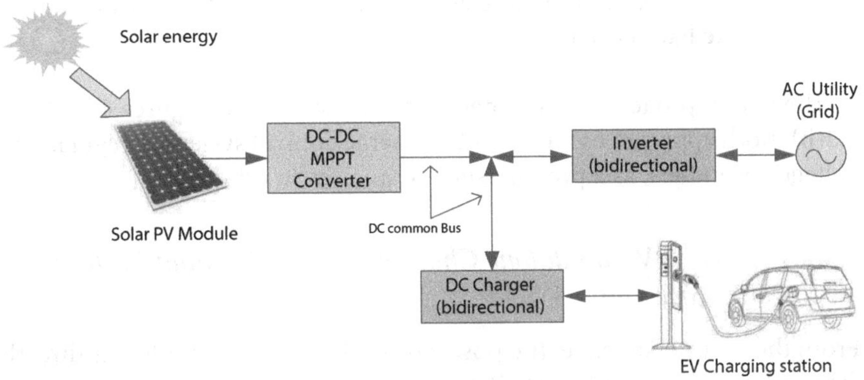

Figure 4.1 A basic block diagram of the solar PV grid-charging system.

A fundamental schematic representation of a solar photovoltaic grid charging system is illustrated in Figure 4.1. All associated components of the solar PV grid charging system were integrated at a convenient point marked as a DC common bus. The range of 200 V–400 V is typically set because the system has different voltages. A DC–DC MPPT converter is based on a boost or buck–boost converter topology, which is implemented to extract the maximum output power from the solar PV module. Furthermore, the PV voltage was constrained to be effectively transferred to the DC bus. The voltage and current were controlled by a DC charger so that they were tuned with the electric vehicles (EVs) being charged at the solar PV grid-charging system. In addition, it has the capacity to flow power bi-directionally. The two applications for installing a bidirectional inverter are that when the supply is extracted using the grid to power the DC bus, it behaves as an AC-to-DC converter (rectification mode). When the bus supply is transferred to the grid, it is configured as a DC-to-AC converter (inversion mode).

4.5.2 Solar PV Standalone Charging Station

In a solar PV standalone charging station, the electric vehicle (EV) is solely charged with the solar PV module without the implication of an electrical grid (utility). This charging station is more efficient because there are fewer stages of power conversion. Regardless, the solar PV module should have sufficient capacity to fulfill the charging requirement for a large number of electric vehicles. Electric vehicles can be charged using two approaches, which are stated below:

I. Solar PV module to electric vehicle (EV) direct charging
II. Solar PV module to electric vehicle (EV) with intermediate battery bank charging.

The two approaches mentioned above illustrated in Figures 4.2(a) and 4.2(b). Both approaches require a DC charge control system to regulate the voltage ensembles, as a prerequisite for an electric vehicle.

4.5.2.1 Solar PV Standalone Charging Station Without Battery Storage Unit (BSU)

From the first perspective, the power from the solar PV module is directly fed to the electric vehicle (EV) for charging, as shown in Figure 4.2(a).

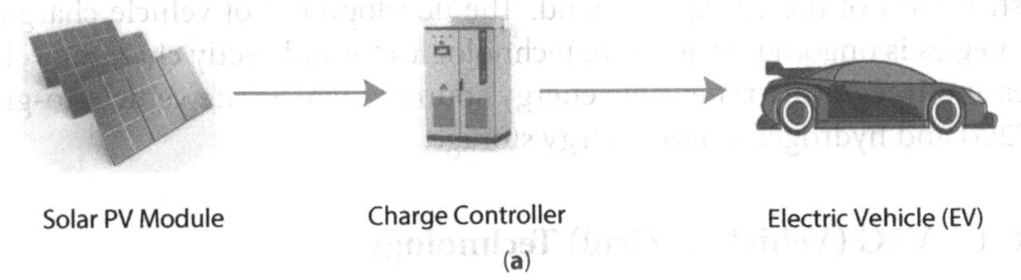

Figure 4.2(a) In solar PV standalone charging station without battery storage unit (BSU).

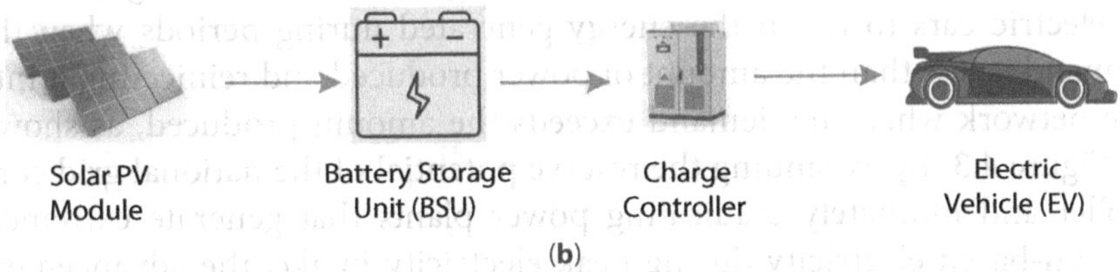

Figure 4.2(b) In solar PV standalone charging station with battery storage unit (BSU).

Its limitation is that the user must compromise with charging when there is an inadequate supply from the solar PV module. Although the setup installation is very simple, as it only requires a single charging controller, this configuration is similar to that of the standalone solar PV system configurations.

4.5.2.2 *Solar PV Standalone Charging Station with Battery Storage Unit (BSU)*

From the second perspective, the intermediate battery storage allows for storing extra power in the battery storage unit (BSU), as shown in Figure 4.2(b), which may be extracted later for charging the electric vehicle when an unavailable power supply is available. The initial investment cost of the battery storage unit (BSU) is high. In some cases, a battery storage unit (BSU) is installed between the solar PV module and electric vehicle to filter the instantaneous variation of the output of the solar PV module.

4.6 Recharging and Innovative Methods

Advancements in electric vehicle technology place additional constraints on distribution and transmission systems. These EVs alter the overall distribution of the energy demand. The development of vehicle charging strategies is ongoing. Innovative technologies for indirectly charging vehicles and integrating renewable energy sources incorporate vehicle-to-grid (V2G) and hydrogen-based energy storage.

4.6.1 V2G (Vehicle to Grid) Technology

The V2G technique is defined as a system that transmits controlled, duplex electrical energy between an electric vehicle and the electrical grid, and is related to electric vehicles. The V2G concept is based on using a battery of electric cars to retain the energy generated during periods when the demand is less than the amount of power produced and reinjecting it into the network when the demand exceeds the amount produced, as shown in Figure 4.3. By expanding the reserve potential of the national grid controller and ultimately eradicating power plants that generate extremely carbon-based electricity during peak electricity intake, the advancement of this structure may aid in regulating electric power flow and integrating renewable energy sources into the grid.

Currently, V2G is a prototype. In the context of EVs, this technology can provide a slew of perks. The dilemma of simultaneously recharging many electric automobiles creates interruptions and several technological challenges for power corporations. However, this approach resolves the issue. V2G can offer network-balanced operations because of its rapid loading and unloading periods. Electric car cells distributed across the system stabilize the frequency of the network. V2G intervention is intended to reduce grid power rates throughout periods of high demand. The authors of [35] demonstrate V2G as a potential strategy for capacity enhancement in distributing infrastructure. V2G may never be sufficient in the context of such statements owing to substantial battery degradation. Additionally, the expense of implementing an intelligent communication network with

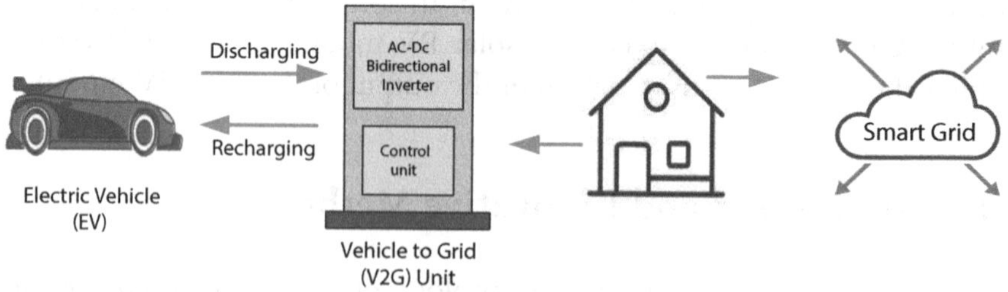

Figure 4.3 V2G (vehicle to grid) technology.

an EV and an electric grid is considerable. Nonetheless, a thorough economic justification in favor of a successful rollout of intelligent adapters and the V2G operation has yet to emerge; it is a subject worth additional investigation.

4.6.2 Hydrogen-Based Energy Storage

Recharging electric vehicles utilizing renewable energy sources can reduce energy prices and GHG emissions and will eventually emerge as a viable means of efficient and environmentally friendly energy consumption [36]. This method requires instantly charging EVs through RESs to automobiles rather than using a hydrogen-based incorporated power-storing framework. Hydrogen is among the most abundant elements on Earth and possesses the maximum mass-based energy density of any fuel. This is because the massive generation of hydrogen has played a critical role in the energy revolution. As shown in Figure 4.4, one or more renewable energy sources are combined in these systems. When the demand for hydrogen and oxygen exceeds that of the natural resources, to generate the additional energy needed, a fuel cell consumes the gases to charge the EVs. Hydrogen is considered a potential future energy vector. Solar hydrogen is produced from renewable energy sources (RES), the most common of which is solar energy.

Since the early 1990s, RESs have been utilized in several projects to produce hydrogen, which is the primary storage medium for renewable energy. Utilizing this resource to power isolated installations, establish an autonomous network, and store extra renewable energy, even if it is not used. As part of this study, hydrogen was found to have a vital effect on mobility.

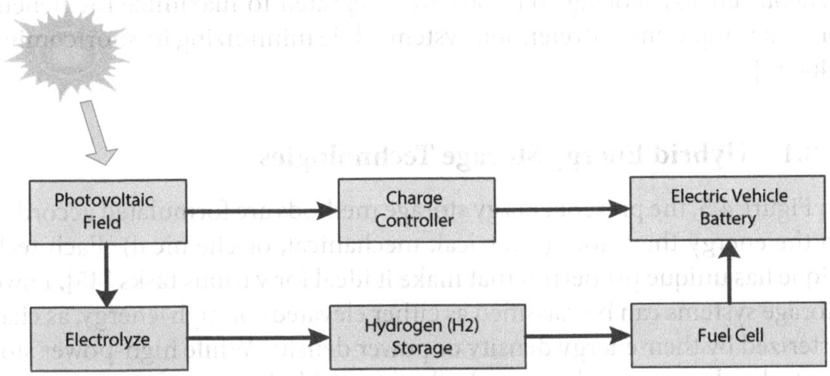

Figure 4.4 Process of producing and storing hydrogen from the sun.

The best aspect of hydrogen is that it can provide a variety of advantages, including congestion management (decline of consumption peak), voltage plan adherence through regulation, resistance to voltage dips, production/absorption of reactive power as needed, and improved energy quality.

4.7 Energy Storage Systems for EV Charging

Energy storage solutions enhance steady operation and improve customer satisfaction concerning EV charging spots, particularly for rapid and ultra-rapid charging devices. To determine their impact on the energy inflow for EV charging, ESS sizing is critical for charging station infrastructure. Table 4.1 lists some studies that have determined the size of energy-storage devices [37].

4.8 Hybrid Energy Storage Technologies to Reduce the Size of the Battery

Energy storage technologies have played a crucial role in the drastic development of mobile energy architectures used in a truly innovative era of hybrid power frameworks. Numerous aspects of commercially accessible power storage innovations are immature [38, 39]. This shows that the dependent type of battery storage must still be fully utilized. In fact, despite extensive research to improve various technologies for storing energy, it is not predicted that a viable technology capable of overcoming all the drawbacks of power storage will be accessible shortly. Therefore, a hybrid energy storage system (HESS) is a viable solution. Two or multiple heterogeneous energy-storing solutions are integrated to maximize the benefits of each unique energy retention system while minimizing its shortcomings [40–44].

4.8.1 Hybrid Energy Storage Technologies

In Figure 4.5, the present energy storage methods are formulated according to the energy they store (electrical, mechanical, or chemical). Each technique has unique properties that make it ideal for various tasks [45]. Power storage systems can be classified as either elevated- or high-energy, as characterized by their energy density or power density. While high-power storage technologies may charge and exhaust quickly, high-energy storage units

Table 4.1 Utilization of energy storage systems for EV charging.

Sr. no.	Charging station type	Energy storage technology used	Result/specification
1.	Ultra-fast charging station	UHSPLB (Ultra high-power superior Lithium Polymer Battery) to facilitate charging of EV batteries sizes varying from16 kWh–75 kWh at a DC level of voltage 800 V	• At a greatest constant charge of 4C and a maximum continuous discharge of 8C, the life cycle is >6,000.
2.	Standalone PHEV charging station	Fuel cell (FC) innovation is intended with the purpose of charging plug-in hybrid electric vehicles (PHEVs). Wind as well as solar resources are used to fuel the recharging outlet, with FC serving energy storage device.	• The H-5000 PEM FC merge with a5-kW rated power is employed. • When compared to a battery bank, using FC for storage is much more cost-effective and has a longer lifespan.
3.	Working place (the usage of a smart or dynamic charging mechanism)	A 10 kWh Li-ion battery is combined with an EV-PV charger to reduce workplace EV charging dependence on the grid by 25%.	• To maintain a long storage lifetime, The discharging depth is restricted to 80% (within the states of charge (SOC) of 10%–90%). • The battery is discharged and charged at a maximum C-rate of 1C, equivalent to a discharging/charging power of Pb max = 10 kW. • Efficiency of charging and draining the battery, comprised of the power converter, is estimated to be 93%, and the grid's energy exchange efficacy is expected to be 95%

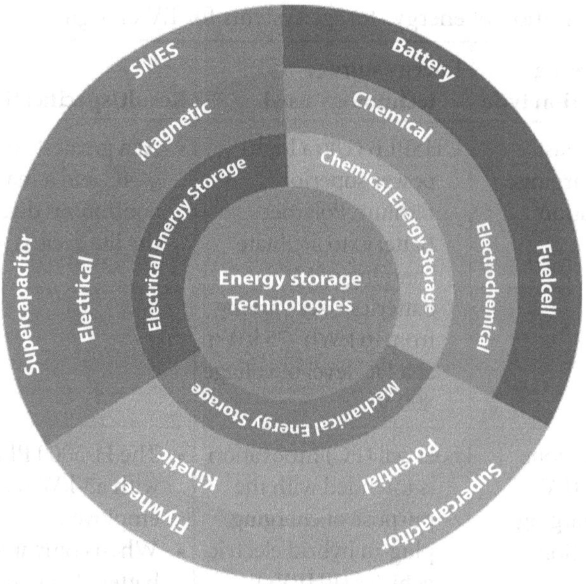

Figure 4.5 Energy storing technologies.

can operate for extended periods but have sluggish dynamics. High-power storage technologies include superconducting magnetic energy storage (SMES), high-power batteries, and supercapacitors. Hydropumped fuel cells, flywheels, high-energy batteries, and compressed air energy retention are examples of high-energy storage techniques. Moreover, the battery energy storage system (BESS) is sufficiently advanced to be featured in both the sections. Lead-acid, nickel-cadmium, and lithium-ion are all forms of high-power storage systems and sodium-sulfur batteries, while high-energy storage systems include zinc-bromine and vanadium redox batteries. HESS combines steep power and high-energy storage device technology. Several HESS can be built technically by combining advanced techniques [46].

4.8.2 Hybrid Energy Storage Challenges

Given the aforementioned possible HESS topologies, more research on adaptable hardware and software is essential to minimize the quantity of energy conserved in mobile systems, autonomous control, and strategies for numerous HESS schemes. Several open issues and obstacles include the following: (i) utilization of modular DC sources to the fullest extent possible using sophisticated switching and control methods [47],

(ii) the adoption of supercapacitors to mitigate pulse burdens and power surges amid transients, (iii) using optimum and adaptive control methods in a single system, temporal coordination, and regeneration energy utilization and [48] (iv) vehicles and infrastructure must work together in cyberspace. Several hybrid energy systems exist, including (i) gadgets and portable systems, such as drones and electric vehicles; (ii) portable microgrid for isolated places; and (iii) aircraft and ships outfitted with powerful radars and pulsed weapons, among others, which will benefit from these techniques.

4.8.3 Challenges in Electric Vehicles

EVs face several obstacles, but the warranty duration of the battery is one of the most significant. Most of the time, this warranty was lower than that of the other components of the vehicle. There insufficient charging stations in this field, which is a problem. The EV battery is designed to last a long period, but degrades over time, just like any other type of battery. Electric vehicle manufacturers provide warranties to safeguard EV owners. As illustrated in Figure 4.6, EV battery life can be severely influenced in a variety of ways [49]. Typically, warranties cover a set period and a certain board range, and they expire when one reaches its threshold. Typically, the warranty ensures that the battery performance does not deteriorate beyond a certain percentage of its warranty period's original charge capacity.

4.9 Battery Management System (BMS)

The battery state of charge (SoC) is monitored consistently to avert overcharging during the plugged-in of an electric vehicle (EV) to being charged then. This monitoring process is managed by a battery management system (BMS). Overdischarging is also managed using a battery management system (BMS), which occurs in the case of the V2G charging system [50].

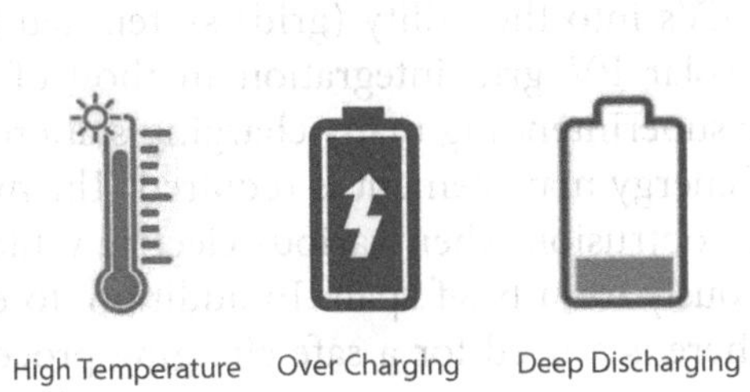

Figure 4.6 Factors that affect the battery life of electric vehicles.

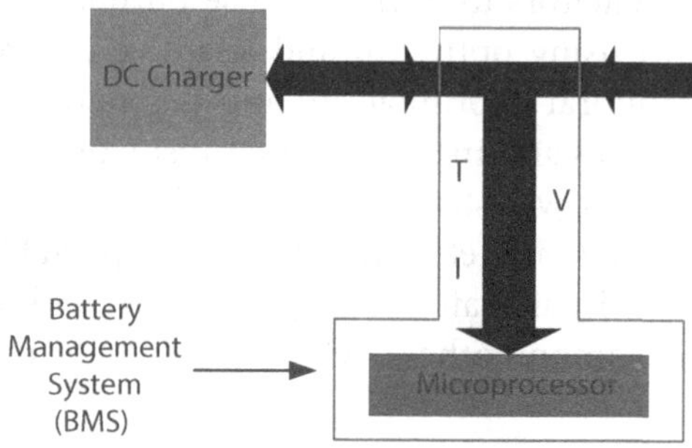

Figure 4.7 Operational key parameters in battery management system (BMS).

The forecasting of the remaining battery strength uses the state of charge (SoC) data, and the output can be applied to determine the distance that the EV can cover. In a solar PV grid-charging system, switching between the battery bank and utility (grid) is controlled by the BMS. In a stand-alone solar PV charging system, the BMS is used to manage the voltage of the battery and decouple the battery and solar PV when generating more than the allowed maximum charging current. This is inadequate for a single battery to operate traction motors. A battery-electric vehicle (EV) is comprised of thousands of single cells connected in a parallel/series combination to establish a battery module. To protect the cells, it is mandatory to continuously monitor the voltage, current, and temperature as the three key operating parameters. The battery management system consists of switches, sensors, and controllers. Figure 4.7 exhibits the position of the BMS in the electric vehicle system and the parameter of its operation [51].

4.10 Conclusion and Future Aspects

This work culminates in a comprehensive study of electric vehicles (EVs), their operation, and the current scenario of solar PV charging methods. The intrusion of EVs into the utility (grid) system led to an increase in demand for the solar PV grid integration method of charging. It was observed that for superintending many charging stations, a more sophisticated system of energy management is required. The main concern is to avert utility (grid) obtrusion when various electric vehicles are interconnected simultaneously for a brief span. In addition, to ensure long-term EV battery life, there is a need for a safe charging procedure that implements BMS (battery management system). There is also a possible area of research to integrate EV charging with different energy sources (other than solar PV), such as wind energy, standby the battery to the smart grid, and fuel cells. However, there may be an increase in capital cost when this integration is implemented, but it could be cost-effective in terms of long-term operating costs. These fascinating topics provide promising prospects for future research in this field.

References

1. Brown, S., Pyke, D., & Steenhof, P. (2010). Electric vehicles: The role and importance of standards in an emerging market. *Energy Policy, 38*(7), 3797-3806.
2. Catenacci, M., Fiorese, G., Verdolini, E., & Bosetti, V. (2015). Going electric: Expert survey on the future of battery technologies for electric vehicles. In *Innovation under Uncertainty*. Edward Elgar Publishing.
3. Romm, J. (2006). The car and fuel of the future. *Energy Policy*, 34(17), 2609-2614.
4. Notter, D. A., Gauch, M., Widmer, R., Wager, P., Stamp, A., Zah, R., & Althaus, H. J. (2010). Contribution of Li-ion batteries to the environmental impact of electric vehicles. *Environ. Sci. Technol.* https://pubs.acs.org/doi/10.1021/es903729a
5. Tanaka N. *Technology Roadmap: Electric and Plug-In Hybrid Electric Vehicles. International Energy Agency*. Technical report; 2011.
6. Block, D., Harrison, J., Brooker, P., Center, F. S., & Dunn, M. D. (2015). *Electric Vehicle Sales for 2014 and Future Projections*. Electric Vehicle Transportation Center.
7. Mariasiu, F. (2012). Energy sources management and future automotive technologies: Environmental impact. *Int. J. Energy Econ. Policy*, 2(4), 342.
8. Herrmann, F., & Rothfuss, F. (2015). Introduction to hybrid electric vehicles, battery electric vehicles, and off-road electric vehicles. In *Advances in Battery Technologies for Electric Vehicles* (pp. 3-16). Woodhead Publishing.
9. Yong, J. Y., Ramachandaramurthy, V. K., Tan, K. M., & Mithulananthan, N. (2015). A review on the state-of-the-art technologies of electric vehicle, its impacts and prospects. *Renew. Sust. Energ. Rev.*, 49, 365-385.
10. Manzetti, S., & Mariasiu, F. (2015). Electric vehicle battery technologies: From present state to future systems. *Renew. Sust. Energ. Rev.*, 51, 1004-1012.
11. Farooqui, S. A., Shees, M. M., Alsharekh, M. F., Alyahya, S., Khan, R. A., Sarwar, A., ... & Khan, S. (2021). Crystal Structure Algorithm (CryStAl) Based Selective Harmonic Elimination Modulation in a Cascaded H-Bridge Multilevel Inverter. *Electronics, 10*(24), 3070.

12. Rao, Z., Wang, S., & Zhang, G. (2011). Simulation and experiment of thermal energy management with phase change material for ageing LiFePO4 power battery. *Energy Convers. Manage.*, 52(12), 3408-3414.
13. Scrosati, B., & Garche, J. (2010). Lithium batteries: Status, prospects and future. *J. Power Sources*, 195(9), 2419-2430.
14. Farooqui, S. A., Khan, R. A., Islam, N., & Ahmed, N. (2021, November). Cuckoo Search Algorithm and Artificial Neural Network-based MPPT: A Comparative Analysis. In *2021 IEEE 8th Uttar Pradesh Section International Conference on Electrical, Electronics and Computer Engineering (UPCON)* (pp. 1-5). IEEE.
15. Mojumder, J. C., Ong, H. C., Chong, W. T., Izadyar, N., & Shamshirband, S. (2017). The intelligent forecasting of the performances in PV/T collectors based on soft computing method. *Renew. Sust. Energ. Rev.*, 72, 1366-1378.
16. Khan, A. I., Khan, R. A., Farooqui, S. A., & Sarfraz, M. (2021). Artificial neural network-based maximum power point tracking method with the improved effectiveness of standalone photovoltaic system. *AI and Machine Learning Paradigms for Health Monitoring System: Intelligent Data Analytics*, 459-470.
17. McBee, K. D. (2017, March). Transformer aging due to high penetrations of PV, EV charging, and energy storage applications. In *2017 Ninth Annual IEEE Green Technologies Conference (GreenTech)* (pp. 163-170). IEEE.
18. Lutsey, N., & Nicholas, M. (2019). Update on electric vehicle costs in the United States through 2030. *The International Council on Clean Transportation*, 2.
19. Zhang, J., Zhang, L., Sun, F., & Wang, Z. (2018). An overview on thermal safety issues of lithium-ion batteries for electric vehicle application. *IEEE Access*, 6, 23848-23863.
20. Khan, R. A., Sabir, B., Sarwar, A., Liu, H. D., & Lin, C. H. (2022). Reptile Search Algorithm (RSA)-Based Selective Harmonic Elimination Technique in Packed E-Cell (PEC-9) Inverter. *Processes, 10*(8), 1615.
21. Kumar, R. R., & Alok, K. (2020). Adoption of electric vehicle: A literature review and prospects for sustainability. *J. Cleaner Prod.*, 253, 119911.
22. Khan, S., Ahmad, A., Ahmad, F., Shafaati Shemami, M., Saad Alam, M., & Khateeb, S. (2018). A comprehensive review on solar powered electric vehicle charging system. *Smart Sci.*, 6(1), 54-79.
23. Whitburn, G. (2012) "Exploring Green Technology, Fundamental Advantages and Disadvantages of Solar Energy."
24. Islam, N., Nat, A., & Khan, R. A. (2022). Recent Technological Advances in Solar Photovoltaic System and Its Applications in Building Integrated Photovoltaic System. *Sustainable Technology and Advanced Computing in Electrical Engineering: Proceedings of ICSTACE 2021*, 625-636.
25. Tsoutsos, T., Frantzeskaki, N., & Gekas, V. (2005). Environmental impacts from the solar energy technologies. *Energy Policy*, 33(3), 289-296.

26. Islam, N., Mohammad, K., Khan, R. A., Sarwer, Z., & Sarwar, A. (2021, May). Symmetrical and Asymmetrical Multilevel Inverter Topologies with Reduced Device Count: A Review. In *2021 2nd International Conference for Emerging Technology (INCET)* (pp. 1-6). IEEE.
27. Naseem, M., Husain, M. A., Minai, A. F., Khan, A. N., Amir, M., Kumar, J. D., & Iqbal, A. (2021). Assessment of Meta-Heuristic and Classical Methods for GMPPT of PV System. *Trans. Electr. Electron. Mater.*, 1-18.
28. Khan, R. A., Farooqui, S. A., Sarwar, M. I., Ahmad, S., Tariq, M., Sarwar, A., ... & Shah Noor Mohamed, A. (2021). Archimedes optimization algorithm based selective harmonic elimination in a cascaded H-bridge multilevel inverter. *Sustainability*, 14(1), 310.
29. Holechek, J.L.; Geli, H.M.E.; Sawalhah, M.N.; Valdez, R. A Global Assessment: Can Renewable Energy Replace Fossil Fuels by 2050? *Sustainability*, 2022, 14, 4792. https://doi.org/10.3390/su14084792
30. Arif, S. M., Lie, T. T., Seet, B. C., Ayyadi, S., & Jensen, K. (2021). Review of electric vehicle technologies, charging methods, standards and optimization techniques. *Electronics*, 10(16), 1910.
31. Li, S., & Zhang, C. (2009, March). Study on battery management system and lithium-ion battery. In *2009 International Conference on Computer and Automation Engineering* (pp. 218-222). IEEE.
32. Khan, R. A., Farooqui, S. A., Khan, M. H., Sarfraz, M., Luqman, M., & Khan, M. (2022). Dust Deposition on PV Module and Its Characteristics. In *The Effects of Dust and Heat on Photovoltaic Modules: Impacts and Solutions* (pp. 59-95). Springer, Cham.
33. Ma, T., & Mohammed, O. A. (2014). Optimal charging of plug-in electric vehicles for a car-park infrastructure. *IEEE Trans. Ind. Appl.*, 50(4), 2323-2330.
34. Lu, L., Han, X., Li, J., Hua, J., & Ouyang, M. (2013). A review on the key issues for lithium-ion battery management in electric vehicles. *J. Power Sources*, 226, 272-288.
35. Chukwu, U. C., & Mahajan, S. M. (2013). V2G parking lot with PV rooftop for capacity enhancement of a distribution system. *IEEE Trans. Sustain. Energy*, 5(1), 119-127.
36. Liang, Y., Wang, H., & Zhao, X. (2022). Analysis of factors affecting economic operation of electric vehicle charging station based on DEMATEL-ISM. *Comput. Ind. Eng.*, *163*, 107818.
37. Mishra, S., Verma, S., Chowdhury, S., Gaur, A., Mohapatra, S., Dwivedi, G., & Verma, P. (2021). A comprehensive review on developments in electric vehicle charging station infrastructure and present scenario of India. *Sustainability*, 13(4), 2396.
38. Martinez, C. M., Hu, X., Cao, D., Velenis, E., Gao, B., & Wellers, M. (2016). Energy management in plug-in hybrid electric vehicles: Recent progress

and a connected vehicles perspective. *IEEE Trans. Veh. Technol.*, 66(6), 4534-4549.

39. Hemmati, R., & Saboori, H. (2016). Emergence of hybrid energy storage systems in renewable energy and transport applications–A review. *Renew. Sust. Energ. Rev.*, 65, 11-23.
40. Hannan, M. A., Azidin, F. A., & Mohamed, A. (2014). Hybrid electric vehicles and their challenges: A review. *Renewable Sustainable Energy Rev.*, 29, 135-150.
41. Rajendran, G., Vaithilingam, C. A., Misron, N., Naidu, K., & Ahmed, M. R. (2021). A comprehensive review on system architecture and international standards for electric vehicle charging stations. *J. Energy Storage*, *42*, 103099.
42. Bhatti, A. R., Salam, Z., Abdul, M. J. B., & Yee, K. P. (2016). A comprehensive overview of electric vehicle charging using renewable energy. *Int. J. Power Electron. Drive Syst.*, 7(1), 114.
43. Anwar, M. B., Muratori, M., Jadun, P., Hale, E., Bush, B., Denholm, P., ... & Podkaminer, K. (2022). Assessing the value of electric vehicle managed charging: a review of methodologies and results. *Energy Environ. Sci.*.
44. Wang, H., Huang, Y., & Khajepour, A. (2018). Cyber-physical control for energy management of off-road vehicles with hybrid energy storage systems. *IEEE/ASME Trans. Mechatron.*, 23(6), 2609-2618.
45. Martinez, C. M., Hu, X., Cao, D., Velenis, E., Gao, B., & Wellers, M. (2016). Energy management in plug-in hybrid electric vehicles: Recent progress and a connected vehicles perspective. *IEEE Trans. Veh. Technol.*, 66(6), 4534-4549.
46. Deng, Z., Yang, L., Cai, Y., Deng, H., & Sun, L. (2016). Online available capacity prediction and state of charge estimation based on advanced data-driven algorithms for lithium iron phosphate battery. *Energy*, 112, 469-480.
47. Castelli Dezza, F., Musolino, V., Piegari, L., & Rizzo, R. (2019). Hybrid battery–supercapacitor system for full electric forklifts. *IET Electr. Syst. Transp.*, 9(1), 16-23.
48. Naseri, F., Farjah, E., & Ghanbari, T. (2016). An efficient regenerative braking system based on battery/supercapacitor for electric, hybrid, and plug-in hybrid electric vehicles with BLDC motor. *IEEE Trans. Veh. Technol.*, 66(5), 3724-3738.
49. Faisal, M., Hannan, M. A., Ker, P. J., Hussain, A., Mansor, M. B., & Blaabjerg, F. (2018). Review of energy storage system technologies in microgrid applications: Issues and challenges. *IEEE Access*, 6, 35143-35164.

50. Raoofi, T., & Yildiz, M. (2023). Comprehensive review of battery state estimation strategies using machine learning for battery Management Systems of Aircraft Propulsion Batteries. *J. Energy Storage*, *59*, 106486.
51. Mousavi, S., Zadehkabir, A., Siavashi, M., & Yang, X. (2023). An improved hybrid thermal management system for prismatic Li-ion batteries integrated with mini-channel and phase change materials. *Appl. Energy*, 334, 120643.

5

A Review of Maximum Power Point Tracking (MPPT) Techniques for Photovoltaic Array Under Mismatch Conditions

Dushyant Sharma[1], Mohd Faisal Jalil[1*], Mohd Shariz Ansari[2] and R. C. Bansal[3,4]

[1]*Department of Electrical Engineering, AMU, Aligarh, Uttar Pradesh, India*
[2]*Department of Electrical & Electronics Engineering, KIET Group of Institutions, Delhi-NCR Ghaziabad, India*
[3]*Department of Electrical Engineering, University of Sharjah, Sharjah, United Arab Emirates*
[4]*Department of Electrical Engineering, University of Pretoria, Pretoria, South Africa*

Abstract

This study provides a well-structured and concise summary of numerous maximum power point tracking (MPPT) approaches applied to photovoltaic (PV) production structures that may be implemented in mismatch conditions and fractional shades. Thus far, a vast collection of schemes, PV modeling procedures, PV array arrangements, and organized structures have been thoroughly examined. However, each strategy has its own benefits and drawbacks. In conclusion, thorough literature research is required to construct a PV generating system (PGS) in partial shade situations. The MPPT algorithms were thoroughly investigated in this study. This study serves as a valuable resource for researchers interested in conducting large-scale PV system studies in the coming days during partially shaded conditions.

***Keywords*:** Maximum power point, global maximum power, PV array

**Corresponding author*: mfaisaljalil@gmail.com

Mohammed Aslam Husain, Md Waseem Ahmad, Farhad Ilahi Bakhsh, P. Sanjeevikumar and Hasmat Malik (eds.) Photovoltaic Systems Technology, (85–102) © 2024 Scrivener Publishing LLC

5.1 Introduction

A nation's ability to produce and distribute power effectively is crucial for its economic prosperity. The goal of India's rural electrification initiative, launched in the 1950s, was to foster economic growth and enhance rural residents' quality of life. The residential, healthcare, agricultural, educational, and business sectors may benefit from the adoption of photovoltaic generating systems (PGS) to create clean power from sunshine. Currently, photovoltaic (PV) technology has the potential to electrify every region of the world. It is very important in PGS to acquire the most power possible from PV panels without being impacted by changes in irradiance throughout the day. However, owing to partial shading, the power output of the PV array is substantially decreased, reducing productivity, increasing design complexity, and increasing price [1–3]. When the irradiance is constant, the PV array characteristics graph has just only maximum power point, which is monitored by either of the well-known maximum power point tracking (MPPT) approaches. However, the PV arrays do not receive consistent solar energy throughout the day. In uneven shade situations, standard MPPT approaches mostly fail because several maxima points arise owing to the occurrence of bypass diodes, which are required to prevent hot spots from forming in PV twines [4, 5]. Many current optimized MPPT algorithms have been proposed to manage multiple maxima under partial-shade situations. This study reassesses different strategies for extracting the optimum amount of electricity from darkened PV systems and is intended to stimulate further research on worldwide PV-based power systems [6]. In previous publications [7], researchers focused mostly on traditional MPPT methods in both irregular and regular irradiance situations. The diverse converter topologies were mostly considered by the author in [8], although a thorough consciousness of all current advances in MPPT for PGS has not yet been completed. This study excludes typical MPPT algorithms and circuit configurations.

The study is organized: *Section 5.1* briefly introduces PV parameters with partial shading. *Section 5.2* provides an evaluation of the various MPPT techniques. These are Perturb and Observe (P&O), Variable Step Size Perturb and Observe, Artificial Neural Network (ANN), Fuzzy Logic Control (FLC), grey wolf optimization (GWO), artificial bee colony-based (ABC), Firefly Algorithm (FA), improved curve tracer, ant colony optimization (ACO), cuckoo search (CS), chaotic search, and the random search method (RSM). Finally, some concluding remarks are presented in *Section 5.3*.

5.2 Evaluation of MPPT Techniques

Researchers can use a variety of MPPT monitoring approaches, and, according to [9, 10], a few are continually being investigated. Several MPPT approaches suitable for consistent irradiance circumstances are listed below. Although some strategies are basic and straightforward to execute, they cannot be used in partial shade, such as Perturb and Observe (P&O) [11–13].

5.2.1 Perturb and Observe (P&O) Technique

Popular strategies include P&O [14, 15]. This procedure is based on a characteristic resulting from the power and voltage of the PV module. The basic idea of this method is that the voltage is purposefully disturbed upwards or downward. The power was then analyzed to analyze previously collected data before the disturbance. Figure 5.1 shows a flowchart representation of the P&O technique.

This will continue in the same way if the interruption results in an increase in power. When the power is reduced, a separate disturbance emerges in the reverse manner. The P & O is Eq. (5.1).

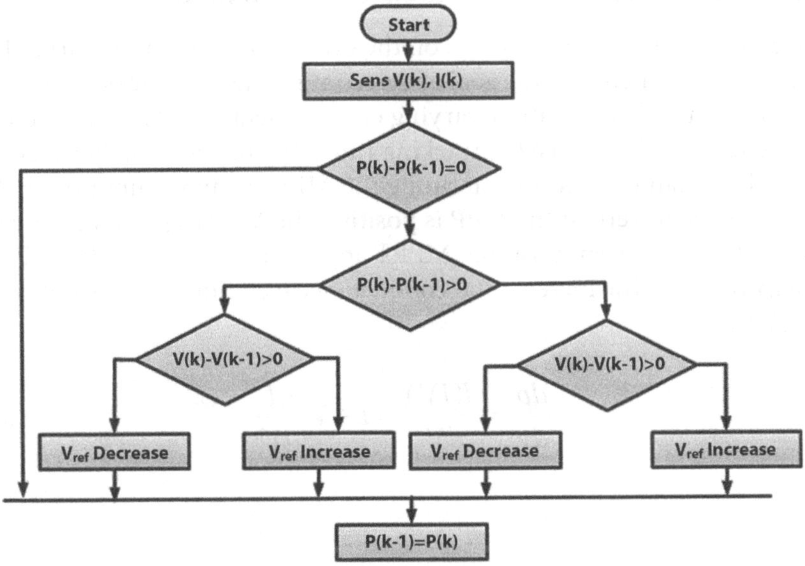

Figure 5.1 Flowchart representation of P&O technique.

$$\frac{\partial P_{PV}}{\partial V_{PV}} = 0 \tag{5.1}$$

where the PV power output and voltage are denoted by P_{PV} and V_{PV}, respectively.

5.2.2 Perturb and Observe Algorithm with Variable Step Magnitude

The P&O strategy was improved using this method, as described in [16, 17]. The reference voltage (Vref) is established at approximately 0.80 Voc at the beginning of the technique, where Voc is the open-circuit potential of the array. It operates in two modes: MPP search and voltage search. The voltage-search phase brings the operational point closer to the reference voltage, whereas the MPP search phase fine-tunes the operating point closer to the MPP. The peak is initially monitored using the revised P&O method. GMPP traceability was used to determine the possibility of partial shade. Unwanted GMPP tracking is eliminated using this strategy. The proposed approach tracks the MPP faster than the P&O approach [18–20].

5.2.3 MPPT Based on Incremental Conductance

The gradual conductance focuses on the evaluation of the P–V curve. This method was created with the goal of overcoming the drawbacks of the P&O scheme. In a wide irradiation-varying environment, the IC seeks to create extra energy and enhance the tracking time. The relationship between dI/dV and I/V can be used to determine the MPP. At maximum power, dP/dV was equal to zero. If the MPP is positive, the MPPT appears on the left side of the most recent position. MPPT appears on the right side if dP/dV is negative [21]. The incremental conductance technique has the following Eq. (5.2).

$$\frac{dp}{dv} = \frac{d(IV)}{dV} = I + V\frac{dI}{dV} \tag{5.2}$$

When dP/dV = 0, MPP is attained.

$$\frac{dP}{dV} > 0 \; then \; V_p < V_{mpp},$$
$$\frac{dP}{dV} = 0 \; then \; V_p = V_{mpp},$$
$$\frac{dP}{dV} \langle 0 \; then \; V_p \rangle V_{mpp},$$

MPP may be located using IC technologies, resulting in lower system costs and power loss. The major drawback of the IC approach is monitoring time, which is comparatively sluggish and not swift. The voltage adjustment factor blames this drawback.

5.2.4 Artificial Neural Network (ANN)-Based MPPT

Bio-inspired algorithms include ANNs which attempt to replicate how the human brain processes information. A general diagram of the operation is shown in Figure 5.2. In this method, a neuron is a processing system that first sequentially evaluates inputs, after which the sum is elaborated using a nonlinear function known as the activation function (AF). Finally, the findings are applied to the next set of neurons [22]. Eq. (5.3) represents the ANN framework, as well as a typical neuron's model.

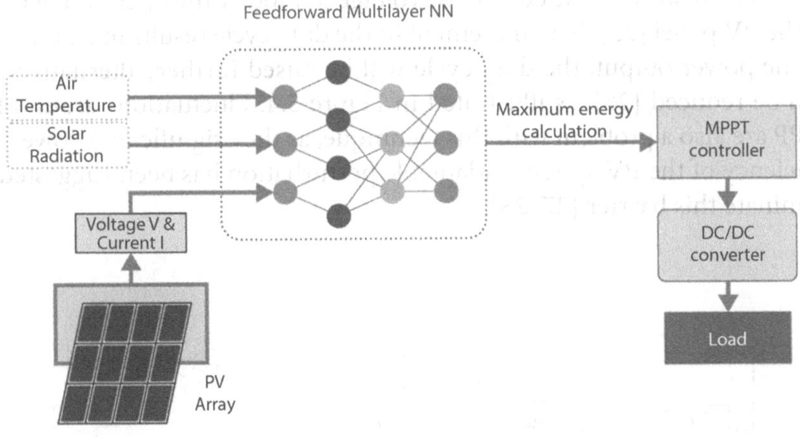

Figure 5.2 Block diagram of an ANN MPPT controller.

$$Z = \sum_{m=1}^{M} \omega_m x_m + \alpha \tag{5.3}$$

where x1, x2,...,xM are the receiving signals. w1, w2,...,wM are the values of the connected synapses, and z is the AF's argument.

5.2.5 The Fuzzy Logic Control (FLC)-Based MPPT

The FLC method is a rule-dependent system that solves nonlinear optimization issues with a high level of resilience (the structure is displayed in Figure 5.3). The Fuzzy technique is chosen for MPPT execution owing to benefits such as simplicity of execution, accessible user interaction, certified validation, and flexible operation. Instead, FLC is rarely used by itself. This is beneficial when coupled with other techniques, such as genetic techniques, artificial neural networks, and different standard approaches [23].

The FLC approach uses three MPP tracking steps for MPP tracking: defuzzification, inference from fuzzy rules, and fuzzification. The fuzzy inference system and fuzzy rule design are steps that determine the peak performance of the system. However, considerable information and expertise are required [24].

5.2.6 Hill Climbing Control-Based MPPT

This approach works based the same principles as the P&O algorithm. The primary distinction between these two algorithms is the alteration of the duty cycle instead of the current or voltage to modify the operational point of the PV panel [25]. If an increment in the duty cycle results in an increase in the power output, the duty cycle will be raised further; alternatively, it will be reduced [26], as illustrated in Figure 5.4. Fluctuations around the MPP are also a problem with this technique, as they significantly lower the efficiency of the PV system. Adaptable perturbation has been suggested to eliminate this barrier [27, 28].

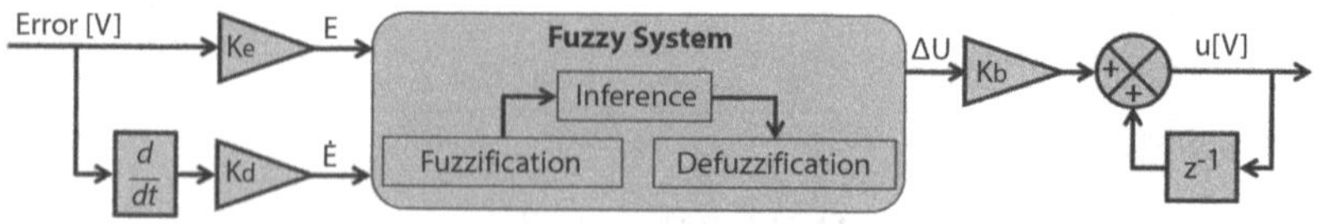

Figure 5.3 FLC structure.

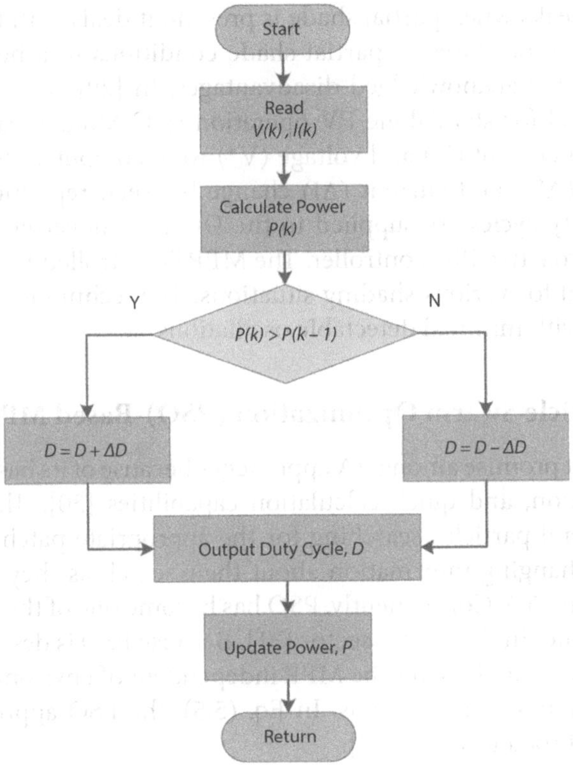

Figure 5.4 Flow diagram of HC.

The suggested approach is utilized to minimize the oscillations that have developed, and thus, the perturbation size is adjusted as follows Eq. (5.8).

$$s = \frac{\Delta P / \Delta d}{P / d} \tag{5.4}$$

Here, s signifies the size of the perturbation step, and P/d signifies how the duty cycle and power values have changed from the two preceding perturbations.

5.2.7 Global Maximum Power Point (GMPP) Technique

The incremental conductance methodology is the origin of the idea for the GMPP method. The tough problem of identifying the global optimum during partial shading is addressed by this technique. The P-V curve will have many peaks when partial shade is present, it deals with the failure of conventional procedures in partial shade conditions in a better manner to overcome the acknowledged disadvantages. In [29], a straightforward GMPP control for standalone PV operation (DC Motor) was presented. The standard current (I*) and voltage (V*) were computed depending on the voltage (ΔV) and current (ΔI) change between repetitions (k). The produced duty cycles are supplied to the DC-DC converter with the aid of a recent proactive PIC controller. The MPPT-controlled DC motor load was evaluated for various shading situations. The technique determines a global peak with minimal detectable oscillations.

5.2.8 Particle Swarm Optimization (PSO)-Based MPPT

PSO has great promise among EA approaches because of its basic structure, cheap execution, and quick calculation capabilities [30]. This approach involves several particles searching for the appropriate patch and collaboratively exchanging information about their search as they do, as illustrated in Figure 5.5. Consequently, PSO has become one of the most widely utilized schemes in the MPPT sector [31]. Because PSO is designed to seek optimization, it can discover the MPP independent of environmental variables for all types of P-V graphs. In Eq. (5.5), the PSO approach can be expressed in broad terms as [32].

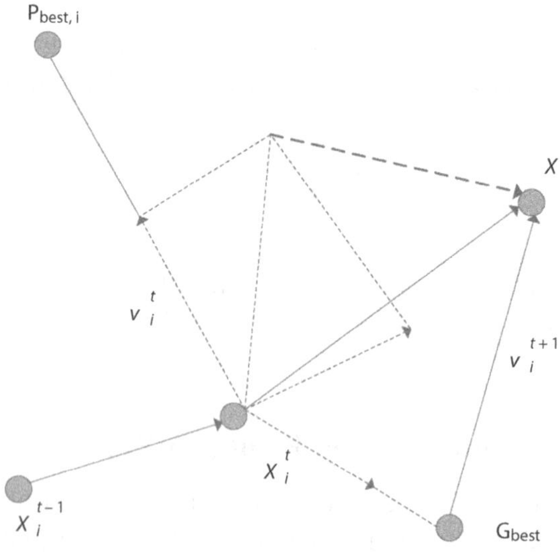

Figure 5.5 PSO particle movement method.

$$x_i^{k+1} = x_i^k + v_i^{k+1} \tag{5.5}$$

where v_i shows the velocity component as determined by Eq. (5.6).

$$v_i^{k+1} = wv_i^k + c_1r_1\left(P_{besti} - x_i^k\right) + c_2r_2\left(G_{besti} - x_i^k\right) \tag{5.6}$$

where the inertia weight is w, G_{best} and P_{best} are the global and personal best locations, and the acceleration parameters are c_1 and c_2, respectively.

5.2.9 Constant Voltage-Based MPPT

The MPPT method for constant voltage is also known as the fractional open-circuit voltage MPPT process. This approach was used in the offline mode. In other words, it is the fraction of MPP to VOC (open-circuit PV module), which is represented in Eqs. (5.7) and (5.8), respectively.

$$V_{REF} = K_V * V_{OC} \tag{5.7}$$

$$\frac{V_{MPP}}{V_{OC}} \cong K_V < 1 \tag{5.8}$$

where VOC is the open-circuit voltage, VMPP is the maximum voltage, and KV denotes a fixed value that varies between 72% and 79% [33, 34].

5.2.10 Constant Current-Based MPPT

The short-circuit current MPPT technique works in the same manner as the open-circuit voltage technique. This approach uses proportionate-to-short-circuit current MPPT [35, 36]. As seen below, the MPPT assumption is proportional to Isc (short circuit current), which is represented in Eqs. (5.9) and (5.10), respectively.

$$I_{REF} = I_{SC} * K_I \tag{5.9}$$

$$\frac{I_{MPP}}{I_{SC}} \cong K_I < 1 \tag{5.10}$$

where, ISC is the short-circuit current, IMPP is the maximum current, and KI is a fixed value, which varies from 78% to 92% [37].

5.2.11 Grey Wolf Optimization (GWO) Algorithm

This metaheuristic technique is heavily influenced by grey wolf-hunting techniques. This approach can emulate both the hunting abilities and leadership hierarchy of grey wolves. Hence, grey wolves are divided into four groups: alpha (α), delta (δ), omega (ω), and beta (β), which are used to accurately imitate the leadership structure [38]. The given Eqs. (5.11) and (5.12) were used to simulate the complete hunting process of grey wolves:

$$\vec{E} = \left|\vec{C}.\overrightarrow{X_P}(t) - \overrightarrow{X_P}(t)\right| \tag{5.11}$$

$$\vec{X}(t+1) = \overrightarrow{X_P}(t) - \vec{F}\vec{E} \tag{5.12}$$

The coefficient vectors are characterized by E, C, and F. The current iteration is denoted as t. The hunting prey's position vector is denoted by Xp, whereas the grey wolf's position vector is denoted by X.

5.2.12 Ant Colony Optimization (ACO)-Based MPPT

A novel optimization approach known as ACO is based on the steps of ants and is used in the following cases: data mining [39], redundancy distribution, and weapon target optimization [40]. To find food, ants usually wander in the N-dimensional territory. They altered the pheromone lay phenomenon (solution grade). The interaction of the ants determines the route to advance from a bad to a reasonable solution [41]. The ACO process of looking for food was modeled and utilized to discover the MPP in PV Systems [42]. Eq. (5.13) gives the likelihood of an ant in an N-dimensional territory containing M colonies as:

$$P_{ij} = \left(\frac{\tau_i^{\alpha} \eta_{ij}^{\beta}}{\sum_{j=1}^{m} \tau_i^{\alpha} \eta_{ij}^{\beta}} \right) \text{when } \eta_i^j > 0, \text{otherwise } P_{ij} > 0 \tag{5.13}$$

where τ_i^j denotes the attraction intensity in area 'I,' 'η_{ij}' means the maximum power sought by the section 'i' to 'j,' where, 'α, β' denotes the influencing variables.

5.2.13 Artificial Bee Colony (ABC) Technique

Bee behavior served as an inspiration for ABC MPPT. In general, bees search for food in various colonies. Two species of bees, the onlooker bee and the employer bee, are deeply involved in the quest for food. A food source is sought by the worker bee when the observer bee is thinking about which source of food to select. This feature was introduced in MPPT to improve tracking speed [43]. Equations (5.14) and (5.15) are the formulas utilized in the ABC MPPT for duty cycle formation and upgrading.

$$d_i = d_{min} + rand[0,1](d_{max} - d_{min}) \tag{5.14}$$

$$newd_i = d_i + \varnothing_i(d_j - d_k) \tag{5.15}$$

where 'd' stands for duty cycle, 'i' for iteration number, $\varnothing_i$ for random integer among [−1, 1], and 'd_i, d_k' for erratically picked duty cycles.

5.2.14 Firefly Algorithm (FA)-Based MPPT

FA is based on the typical behavior of fireflies with varying light intensities [44]. Regardless of MPPT, the firefly technique is used for current relay synchronization [45] and supports vector regression prediction applications [46, 47]. In [47], a revised firefly technique was used to solve the dimensional issue after detecting capacity in the FA. The relative brightness of fireflies is proportional to their distance. The fireflies with less intelligence will go towards illumination with greater intensity.

5.2.15 Curve Tracer MPPT

Software-based MPPT techniques primarily use curved tracers [48]. It is divided into different phases, one of which is primarily the boost converter. The other incorporates an improved hysteretic self-discipline circuit and a duty-modulated load resistance (DMLR). Changes are made to the panel's ideal position using a boost converter for loading. The improved curve tracer includes two working modes: modulated resistance (MR) and fixed resistance (FR) mode [49]. In the MR states, the equivalent resistance of the load switch is controlled by the DMLR. However, in the FR state, the curve tracer functions as a traditional boost V-I arc tracer. The enhanced curve tracer has a straightforward structure, low implementation costs, and no trace restrictions.

5.2.16 Cuckoo Search (CS)-Based MPPT

The CS approach is a bio-inspired parasitic reproductive system used by cuckoo birds [50]. The levy-flying mechanism is used in CS to describe the procedures involved in finding a nest. The step values for the levy distribution function are obtained from a levy flight, which is primarily a random process, as shown in Figure 5.6. CS step sizes are more significant than those of conventional PSOs owing to the Levy's flying mechanism. This has hastened the convergence. The step size decreases as the particles approach the MPP, eventually reaching zero.

5.2.17 Chaotic Search-Based MPPT

Chaos is described as the unpredictability that happens to own a system's sensibleness to a slight change in the initial circumstances. Consequently, the search for MPPT optimization in PGS is based on this characteristic in [51]. The optimizing parameter in the PGS is the voltage and the fitness function is the power. There are two types of optimization methods: single-carrier and multiple-carrier. Single-carrier searching is usually impractical because it takes significantly longer to reach the MPP, but various carriers can enhance the accuracy, performance, and system reliability. The search region shrinks and eventually stops when it reaches the threshold level, allowing for faster convergence.

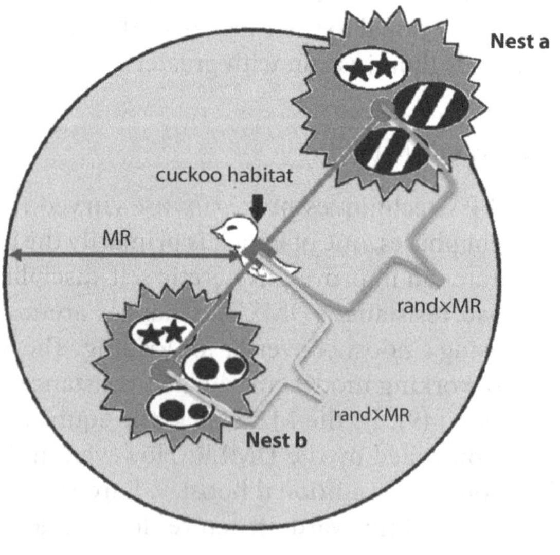

Figure 5.6 Cuckoo optimization algorithm.

5.2.18 Random Search Method (RSM)-Based MPPT

Anderson proposed a unique strategy validated by Rastrigin, who created a mathematical formula for RSM. It is was later used to retrieve hyper-parameters in [52]. RSM typically selects random values within a given consolidation and then searches for global optimization. The duty cycle parameters were created randomly in the MPPT technique and modified until the MPP was obtained. Equations (5.16) and (5.17) used in the implementation are as follows:

$$d^{k+1} = d^k + \lambda u \tag{5.16}$$

$$f^{k+1} >= f\left(d^{k+1}\right) \tag{5.17}$$

Where 'k' represents the number of iterations, 'λ' represents the scaling factor, 'd' represents the duty cycle, and 'u' represents a random value [−1, 1].

5.3 Conclusion

This article provides a brief overview of all the most widely used MPPT techniques. It discusses MPPT optimization strategies that primarily focus on PGS with partially shaded circumstances. In addition to the new MPPT techniques, this evaluation included several recent hybrid approaches. The use of a PV modeling technique under partial shade for better monitoring and easier detection of power peaks was presented. Several new PV circuit configurations have been designed to improve the tracking efficiency of PV system. The advantages and disadvantages of various optimization strategies were examined to select an appropriate MPPT for partial shading. It is challenging to determine which method is the best among the different methods presented in this study. The system's purpose, hardware accessibility, expense, convergence rate, accuracy, and dependability influence the MPPT selection. Owing to the importance of MPPT in partial shade conditions, there is sufficient space for research to discover an appropriate MPPT that can increase the PGS output efficiency. This study is intended to be a useful tool for all PV researchers as well, and to excel in creating efficient, clean, and renewable resources for humanity.

References

1. Subudhi, B., Pradhan, R.: A comparative study on maximum power point tracking techniques for photovoltaic power systems. *IEEE Trans. Sustain. Energy*, 4, 89–98 (2013). https://doi.org/10.1109/TSTE.2012.2202294.
2. Husain, Mohammed Aslam; Jain, Abhinandan; Tariq, Abu; Iqbal, Atif (2019). Fast and precise global maximum power point tracking techniques for photovoltaic system. *IET Renewable Power Gener.*, 13(14), 2569–2579. doi:10.1049/iet-rpg.2019.0244
3. Nasiruddin, I., Khatoon, S., Jalil, M.F., Bansal, R.C.: Shade diffusion of partial shaded PV array by using odd-even structure. *Sol. Energy.*, 181, 519–529 (2019). https://doi.org/10.1016/j.solener.2019.01.076.
4. Jalil, M.F., Khatoon, S., Nasiruddin, I., Bansal, R.C.: An Improved Feasibility Analysis of Photovoltaic Array Configurations and Reconfiguration Under Partial Shading Conditions. *Electr. Power Components Syst.*, 48, 1077–1089 (2020). https://doi.org/10.1080/15325008.2020.1821842.
5. M. Naseem *et al.*, "A Spider Monkey Optimization Based Global Maximum Power Point Tracking Technique for Photovoltaic Systems," *2022 2nd International Conference on Emerging Frontiers in Electrical and Electronic Technologies (ICEFEET)*, Patna, India, 2022, pp. 1-6, doi: 10.1109/ICEFEET51821.2022.9847903.
6. M. Naseem; Mohammed Aslam Husain; Ahmad Faiz Minai; Ahmad Neyaz Khan; Mohd Amir; J. Dinesh Kumar; Arif Iqbal; (2021). Assessment of Meta-Heuristic and Classical Methods for GMPPT of PV System. *Trans. Electr. Electron. Mater.*, (), –. doi:10.1007/s42341-021-00306-3
7. D. Sharma, M. F. Jalil, M. S. Ansari, and R. C. Bansal, "A review of PV array reconfiguration techniques for maximum power extraction under partial shading conditions," *Optik (Stuttg)*, p. 170559, Jan. 2023, doi: 10.1016/j.ijleo.2023.170559.
8. Jalil, M.F., Saxena, R., Ansari, M.S., Ali, N.: Reconfiguration of photo voltaic arrays under partial shading conditions. In: *Second International Innovative Applications of Computational Intelligence on Power, Energy and Controls with their Impact on Humanity (CIPECH)*. pp. 193–200. IEEE (2016). https://doi.org/10.1109/CIPECH.2016.7918765.
9. N. U. Islam Wani, F. I. Bakhsh, P. Choudekar and Ruchira, "Active Power Control of Grid Connected SPV Plant Based Microgrid Using Active Power Regulating Scheme," *2021 Emerging Trends in Industry 4.0 (ETI 4.0)*, OP Jindal University, Raigarh, Chhattisgarh, India, pp. 1-8, 19 - 21 May 2021. doi: 10.1109/ETI4.051663.2021.9619213
10. N. Sharma, F. I. Bakhsh, S. Mehta, "Efficiency Enhancement of a Solar Power Plant Using Maximum Power Point Tracking Techniques" *IEEE International Conference on Computational and Characterization Techniques in Engineering & Sciences (CCTES-18)*, Integral University, Lucknow, India, pp. 106-111, 14-15 September, 2018. doi: 10.1109/CCTES.2018.8674146

11. Mohanty, S., Subudhi, B., Ray, P.K.: A new MPPT design using grey Wolf optimization technique for photovoltaic system under partial shading conditions. *IEEE Trans. Sustain. Energy*, 7, 181–188 (2016). https://doi.org/10.1109/TSTE.2015.2482120.
12. A. Kumar, M. Andleeb, F. I. Bakhsh, "Design and Analysis of Solar PV Rooftop in Motihari," *J. Phys. Conf. Ser.*, vol. 1817, no. 1, pp. 1-14, March 2021. doi:10.1088/1742-6596/1817/1/012019
13. Jalil, M.F., Khatoon, S., Nasiruddin, I.: Improved Design Analysis of Photovoltaic Modules and Arrays Under Varying Conditions. In: *2018 3rd International Innovative Applications of Computational Intelligence on Power, Energy and Controls with their Impact on Humanity (CIPECH)*. pp. 68–75. IEEE (2018). https://doi.org/10.1109/CIPECH.2018.8724237.
14. Hua, C., Lin, J., Shen, C.: Implementation of a DSP-controlled photovoltaic system with peak power tracking. *IEEE Trans. Ind. Electron.*, 45, 99–107 (1998). https://doi.org/10.1109/41.661310.
15. Salameh, Z., Taylor, D.: Step-up maximum power point tracker for photovoltaic arrays. *Sol. Energy*, 44, 57–61 (1990). https://doi.org/10.1016/0038-092X(90)90027-A.
16. Belhachat, F., Larbes, C.: A review of global maximum power point tracking techniques of photovoltaic system under partial shading conditions. *Renew. Sustain. Energy Rev.*, 92, 513–553 (2018). https://doi.org/10.1016/j.rser.2018.04.094.
17. Femia, N., Petrone, G., Spagnuolo, G., Vitelli, M.: Optimization of perturb and observe maximum power point tracking method. *IEEE Trans. Power Electron.*, 20, 963–973 (2005). https://doi.org/10.1109/TPEL.2005.850975.
18. Esram, T., Chapman, P.L.: Comparison of photovoltaic array maximum power point tracking techniques. *IEEE Trans. Energy Convers.*, 22, 439–449 (2007). https://doi.org/10.1109/TEC.2006.874230.
19. Kashani, M.G., Mobarrez, M., Bhattacharya, S.: Variable interleaving technique for photovoltaic cascaded DC-DC converters. In: *IECON Proceedings (Industrial Electronics Conference)*. pp. 5612–5617 (2014). https://doi.org/10.1109/IECON.2014.7049359.
20. Kamran, M., Mudassar, M., Fazal, M.R., Asghar, M.U., Bilal, M., Asghar, R.: Implementation of improved Perturb & Observe MPPT technique with confined search space for standalone photovoltaic system. *J. King Saud Univ. - Eng. Sci.*, 32, 432–441 (2018). https://doi.org/10.1016/j.jksues.2018.04.006.
21. D. P. Hohm and M. E. Ropp. Comparative Study of Maximum Power Point Tracking Algorithms Using an Experimental, Programmable, Maximum Power Point Tracking Test Bed, *Proc. Photovoltaic Specialist Conference*, 2000, pp. 1699–1702.
22. Bhatnagar, P., Nema, R.K.: Maximum power point tracking control techniques: State-of-the-art in photovoltaic applications. *Renew. Sustain. Energy Rev.* 23, 224–241 (2013). https://doi.org/10.1016/j.rser.2013.02.011.

23. Bounechba, H., Bouzid, A., Nabti, K., Benalla, H.: Comparison of perturb & observe and fuzzy logic in maximum power point tracker for PV systems. *Energy Procedia,* 50, 677–684 (2014). https://doi.org/10.1016/j.egypro.2014.06.083.
24. Mohd Zainuri, M.A.A., Mohd Radzi, M.A., Soh, A.C., Rahim, N.A.: Development of adaptive perturb and observe-fuzzy control maximum power point tracking for photovoltaic boost dc-dc converter. *IET Renew. Power Gener.*, 8, 183–194 (2014). https://doi.org/10.1049/iet-rpg.2012.0362.
25. K Ishaque, Z. Salam, A review of maximum power point tracking techniques of PV system for uniform insolation and partial shading condition, *Renew. Sustain. Energy Rev.*, 19 (2013) 475–488.
26. F Liu, Y Kang, Y Zhang, S Duan, Comparison of P&O and hill climbing MPPT methods for grid-connected PV converter, in: *Proceedings of the 3rd IEEE Conference on Industrial Electronics and Applications ICIEA*, IEEE, 2008, pp. 804–807.
27. ML Chiang, CC Hua, JR Lin, Direct power control for distributed PV power system, In: *Proceedings of the Power Conversion Conference PCC*. IEEE (2002) 311–315.
28. W Xiao, WG. Dunford, A modified adaptive hill climbing MPPT method for photovoltaic power systems, in: *35th Annual Power Electronics Specialists' Conference PESC*, IEEE, 2004, pp. 1957–1963.
29. Bouilouta Mellit A, Kalogirou SA. New MPPT method for stand-alone photovoltaic systems operating under partially shaded conditions. *Energy,* 2013;55:1172–85.
30 Chai, L.G.K., Gopal, L., Juwono, F.H., Chiong, C.W.R., Ling, H.-C., Basuki, T.A.: A novel global MPPT technique using improved PS-FW algorithm for PV system under partial shading conditions. *Energy Convers. Manag.* 246, 114639 (2021). https://doi.org/10.1016/j.enconman.2021.114639.
31. Ahmed, J., Salam, Z.: An Enhanced Adaptive P & O MPPT for Fast and Efficient Tracking Under Varying Environmental Conditions. *IEEE Trans. Sustain. Energy.* 9, 1487–1496 (2018). https://doi.org/10.1109/TSTE.2018.2791968.
32. M. Nisha, M. Andleeb, F. I. Bakhsh, "Effect of Partial Shading on a PV Array and Its Maximum Power Point Tracking Using Particle Swarm Optimization," *J. Phys. Conf. Ser.*, vol. 1817, no. 1, pp. 1-14, March 2021. doi:10.1088/1742-6596/1817/1/012025
33. L Van Der Merwe, GJ Van Der Merwe, Maximum power point tracking-implementation strategies [for PV power systems]. In *IEEE International Symposium on Industrial Electronics, Proceedings* 1 (1998) 214–217.
34. MA El Ela, JA Roger, Optimization of the function of a photovoltaic array using a feedback control system, *Solar Cells,* 13 (1984) 107–119.
35. X Di, M Yundong, C Qianhong, A global maximum power point tracking method based on interval short-circuit current, in: *2014 16th European Conference on Power Electronics and Application*, 2014, pp. 1–8.

36. ZM Salameh, F Dagher, WA Lynch, Step-down maximum power point tracker for photovoltaic systems, *Sol. Energy*, 46 (1991) 279–282.
37. T Esram, PL. Chapman, Comparison of photovoltaic array maximum power point tracking techniques, *IEEE Trans. Energy Convers.*, 22 (2007) 439–449.
38. Shi, J.-Y., Zhang, D.-Y., Ling, L.-T., Xue, F., Li, Y.-J., Qin, Z.-J., Yang, T.: Dual-algorithm maximum power point tracking control method for photovoltaic systems based on grey wolf optimization and golden-section optimization. *J. Power Electron.*, 18, 841–852 (2018). https://doi.org/10.6113/JPE.2018.18.3.841.
39. Lee, Z.J., Lee, C.Y., Su, S.F.: An immunity-based ant colony optimization algorithm for solving weapon-target assignment problem. *Appl. Soft Comput. J.*, 2, 39–47 (2002). https://doi.org/10.1016/S1568-4946(02)00027-3.
40. Liang, Y.C., Smith, A.E.: An ant colony optimization algorithm for the redundancy allocation problem (RAP). *IEEE Trans. Reliab.*, 53, 417–423 (2004). https://doi.org/10.1109/TR.2004.832816.
41. Wani N.U.I., F. I. Bakhsh, Choudekars P., Ruchira (2022) Intelligent Modelling and Analysis of P–Q Control Technique for SPV Plant Supplying Power to Grid. In: Malik H., Ahmad M.W., Kothari D. (eds) *Intelligent Data Analytics for Power and Energy Systems*. Lecture Notes in Electrical Engineering, vol. 802. Springer, Singapore. https://doi.org/10.1007/978-981-16-6081-8_7
42. Jiang, L.L., Maskell, D.L., Patra, J.C.: A novel ant colony optimization-based maximum power point tracking for photovoltaic systems under partially shaded conditions. *Energy Build.*, 58, 227–236 (2013). https://doi.org/10.1016/j.enbuild.2012.12.001.
43. H. Malik, A. Iqbal, P. Joshi, S. Agrawal and F. I. Bakhsh, "Metaheuristic and Evolutionary Computation: Algorithms and Applications," *Studies in Computational Intelligence*, vol. 916, Springer, Singapore, 2021. ISBN 978-981-15-7571-6
44. Farzaneh, J., Keypour, R. & Khanesar, M.A. A New Maximum Power Point Tracking Based on Modified Firefly Algorithm for PV System Under Partial Shading Conditions. *Technol. Econ. Smart Grids Sustain. Energy*, 3, 9 (2018). https://doi.org/10.1007/s40866-018-0048-7 Link: https://link.springer.com/article/10.1007/s40866-018-0048-7
45. Huang, Y.-P., Chen, X., Ye, C.-E.: Research Article A Hybrid Maximum Power Point Tracking Approach for Photovoltaic Systems under Partial Shading Conditions Using a Modified Genetic Algorithm and the Firefly Algorithm. *Int. J. Photoenergy*, 2018, 1–13 (2018). https://doi.org/10.1155/2018/7598653.
46. Sundareswaran, K., Peddapati, S., Palani, S.: MPPT of PV systems under partial shaded conditions through a colony of flashing fireflies. *IEEE Trans. Energy Convers.*, 29, 463–472 (2014). https://doi.org/10.1109/TEC.2014.2298237.
47. Verma, O.P., Aggarwal, D., Patodi, T.: Opposition and dimensional based modified firefly algorithm. *Expert Syst. Appl.*, 44, 168–176 (2016). https://doi.org/10.1016/j.eswa.2015.08.054.

48. Ramaprabha, R., Jubair, S.H., Suhas, K., Lokesh, A.: Development of improved curve tracer for photovoltaic system under partial shaded conditions. *2014 Int. Conf. Circuits, Power Comput. Technol. ICCPCT 2014.* 585–589 (2014). https://doi.org/10.1109/ICCPCT.2014.7054857.
49. Sundareswaran, K., Peddapati, S., Palani, S.: Application of random search method for maximum power point tracking in partially shaded photovoltaic systems. *IET Renew. Power Gener.*, 8, 670–678 (2014). https://doi.org/10.1049/ iet-rpg.2013.0234.
50. Ahmed, J., Salam, Z.: A Maximum Power Point Tracking (MPPT) for PV system using Cuckoo Search with partial shading capability. *Appl. Energy*, 119, 118–130 (2014). https://doi.org/10.1016/j.apenergy.2013.12.062.
51. Zhou, L., Chen, Y., Guo, K., Jia, F.: New approach for MPPT control of photovoltaic system with mutative-scale dual-carrier chaotic search. *IEEE Trans. Power Electron.*, 26, 1038–1048 (2011). https://doi.org/10.1109/TPEL.2010.2078519.
52. Ameryan, Mahya, Mohammad Reza Akbarzadeh Totonchi and Seyyed Javad Seyyed Mahdavi. "Clustering based on Cuckoo Optimization Algorithm." *2014 Iranian Conference on Intelligent Systems (ICIS)* (2014): 1-6.

6

Metaheuristic Techniques for Power Extraction from PV-Based Hybrid Renewable Energy Sources (HRESs)

Akhlaque Ahmad Khan* and Ahmad Faiz Minai†

EED, Integral University, Lucknow, U.P., India

Abstract

The scope of power generation using renewable resources is increasing daily. Currently, the decreasing cost and high efficiency of solar panels have attracted users to install solar-based hybrid renewable plants for commercial and domestic use. Governments have played a significant role in promoting solar energy usage for power production by implementing a variety of user-friendly programs, policies, and tariffs. Furthermore, there are two main types of electricity generation systems that use solar energy, i.e., solar thermal-based plants and solar photovoltaic-based plants. There are two types of solar PV-based power plants: stand-alone (off grid) PV-based HRE systems and grid-connected PV-based HRE systems. These systems are very attractive to consumers because they can be easily installed on the roofs of commercial or non-commercial buildings according to their rating. One of the advantages of these types of rooftop systems is that they do not require extra land, and thus, the system is also cost effective. In this chapter, power extraction from SPV-based plants is discussed. In this era of automation, there is a need for controllers for every system and to optimize the components of the system/resource to increase the efficiency of the overall plant. Currently, some metaheuristic optimization techniques are used to control and operate the plant in a manner that allows the system to become more efficient, reliable, and cost-effective. In this chapter, these metaheuristic optimization techniques are discussed for all types of solar PV-based HRE systems. Finally, a comparison was made of some of the best-suited optimization techniques for power extraction

**Corresponding author*: akhlaque@iul.ac.in
†*Corresponding author*: fzminai@gmail.com

Mohammed Aslam Husain, Md Waseem Ahmad, Farhad Ilahi Bakhsh, P. Sanjeevikumar and Hasmat Malik (eds.) Photovoltaic Systems Technology, (103–122) © 2024 Scrivener Publishing LLC

from solar PV-based HRE systems. Its advantages and limitations are also mentioned for consumers and better understanding of the readers.

***Keywords*:** Metaheuristic optimization techniques, solar PV system, electricity generation, MPPT, battery storage, control and protection strategies

Abbreviation

HRES	Hybrid Renewable Energy System
PV	Photovoltaic
UN	United Nation
SDG	Sustainable Development Goal
GW	Giga Watt
MPPT	Maximum Power Point Tracking
PS	Partial Shadowing
AI	Artificial Intelligence
BI	Bio Inspired
P&O	Perturb and Observe
INC	Incremental Conductance
PSC	Partial Shading Effect
EA	Evolutionary Algorithm
GMPP	Global Maximum Power Point
ANN	Artificial Neural Networks
FLC	Fuzzy Logic Controller
GA	Genetic Algorithms
GWO	Grey Wolf Optimization
ABC	Artificial Bee Colony
ACO	Ant Colony Optimization
FFA	Fire Fly Algorithm
DE	Differential Algorithm
BA	Bat Algorithm
AFLC	Adaptive Fuzzy Logic Controller

6.1 Introduction

The cornerstone of the endeavor to accomplish the goals outlined in the Paris Agreement and the UN SDGs, particularly those pertaining to the battle against climate change and universal access to energy, is renewable energy, which is ecologically beneficial and inexhaustible. The key to achieving these goals is using hybrid generating facilities, which combine two renewable energy sources, solar and wind, to assure a steadier supply [1, 2].

The energy production industry has relied heavily on renewable energy sources over the past 10 years to reduce greenhouse gas emissions. Energy from renewable energy generation is available to all types of energy consumers [3]. Solar energy has the potential to be the finest alternative source for supplying energy, and this has led to its popularity [2, 4]. According to statistics, the solar energy industry is expanding rapidly. Globally, 227 GW of PV plants were installed as of the end of 2015, and a further 50 GW was predicted to be added in 2016. By 2021, it is anticipated that 600 GW of PV plants will be installed globally [2]. The effectiveness of energy conversion in PV systems is between 6% and 20%, which is still underwhelming [2, 3]. Several studies [5, 6] have attempted to identify remedies for this issue to enhance the conversion rate. The power generated by solar PV systems varies with climate due to the unpredictability of meteorological circumstances [7]. Additionally, owing to the characteristics of the atmosphere and obstructing clouds, solar irradiation changes significantly before it hits the PV array. An immediate shift in the peak power point results from this change in these parameters.

An MPPT controller is essential for PV systems since it improves their efficiency [8]. A solar PV system with MPPT was initially implemented in spacecraft in 1968 [9]. Since then, there has been a major increase in reliability, accuracy, tracking ability, effectiveness, and convenience of MPPT controllers. The most efficient MPPT algorithm reacts rapidly to changes in the output power with low-frequency oscillations around the power-phase point. Numerous MPPT control systems have been used and are currently accessible [10]. These regulators were used to follow a set MPP when the light was always the same, or to adapt to a PV system with more than one MPP. Although each PV system ought to utilize a specific MPP, doing so should not require an elaborate procedure. Odd events can occur most of the time. Partial shadowing (PS) is one of the most significant factors that slightly affects the amount of electricity that can be collected from any PV system [11]. Both the GMPP and MMPPP rely on PS as a major aspect of their creation. The efficiency of standard MPPT algorithms is low in PS [10].

According to literature, MPPT controllers can be categorized in several ways. Figure 6.1 shows the typical breakdown of MPPT into its component parts: state-of-the-art (soft computing) methods, conventional methods, and hybrid methods. Conventional MPPT techniques have been extensively employed because of their long history and applications [12].

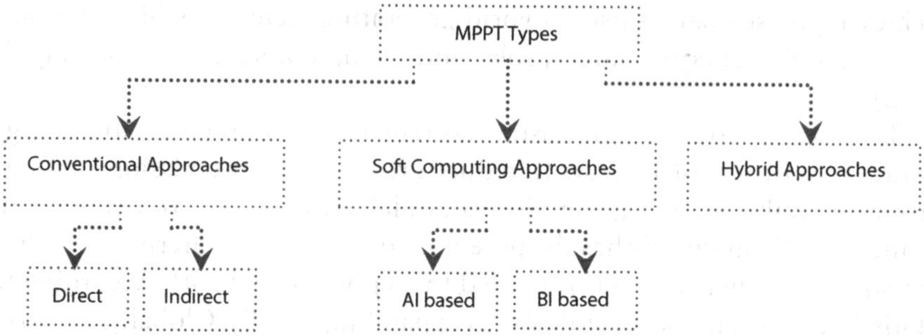

Figure 6.1 Classification of MPPT approaches.

Their main advantages include their high flexibility and usability. The only MPP that can be tracked using these techniques is the exposure to uniform radiation [13]. Names like "artificial intelligence (AI)," "soft computing," and "bio-inspired (BI)" were employed to characterize modern MPPT methods [14]. Despite being rather complex, these approaches are better at tracking than the conventional approaches.

P&O, INC, PC, constant voltage, and constant current are the most commonly used MPPT methods [15, 16]. However, these methods have several drawbacks, making it impossible for them to function in PSCs. Conventional algorithms struggle to identify the GMPP under PSCs; even when they do, they take a lot of time, have poor tracking precision, and exhibit high oscillations. This implies that effective algorithms are required to extract the optimal GMPP [17, 18].

In contrast to traditional approaches, MPPT methods are more sophisticated. Soft computing approaches are excellent choices for PSCs owing to their flexibility, resilience, and dependability. The working premise is mostly based on the evolutionary algorithm (EA) principle and other AI approaches. The most popular modern techniques include artificial neural networks (ANN), fuzzy logic controllers (FLC), genetic algorithms (GA), grey wolf optimization (GWO), artificial bee colonies (ABC), ant colony optimization (ACO), firefly algorithms (FFA), differential algorithms (DE), and bat algorithms (BA). Their flexibility and capacity to address nonlinear problems are key benefits [16]. The best solution or very effective multipeak GMPPT can be delivered by them. Compared with conventional techniques, these algorithms exhibit better tracking performance [19].

Hybrid techniques are an additional type of MPPT approach. Hybrid MPPT methods are frequently used to overcome the limitations of single-method approaches and produce superior outcomes. To address the shortcomings of both the conventional and advanced approaches, hybrid

MPPT methods have been proposed. Combining two conventional techniques, a soft computing-based algorithm with a conventional algorithm, or two soft computing-based algorithms are examples of hybrid approaches [20].

Tracking performance, quickness, correctness, affordability, efficiency, and intricacy are a few of the common parameters used to assess whether the MPPT technique can meet the intended application. Other criteria, such as fluctuations around the MPP, surveillance on PCs and in usual settings, system freedom, and other factors, are also crucial when choosing the optimum technique and need to be considered in addition to these characteristics. The most widely used MPPT techniques have been reviewed in this chapter.

6.2 Hybrid Renewable Energy Systems

HRES combine more than one renewable source to provide a reliable and continuous power supply. In other words, a hybrid energy system is a mix of numerous (two or more) energy sources with suitable energy conversion technologies coupled to supply electricity to local loads/grids. There is no single standard or structure because it falls under the category of a dispersed generating system. It benefits the environment, relieves transmission and distribution congestion, increases system dependability, enhances power quality, and boosts the overall efficiency. It also reduces line and transformer losses [3].

6.2.1 Types of Hybrid Renewable Energy Systems

In addition to hybrid systems utilizing diesel generators, the most popular hybrid power configurations are solar thermal/biomass, photovoltaic/wind, photovoltaic/hydro, and hydro/wind. Based on the uses of the end users, two types of hybrid renewable energy systems were created: 1) a grid-connected HRE system, and 2) a stand-alone or off-grid HRE system [21].

6.2.1.1 Grid-Connected HRE System

A solar array and a wind turbine were combined to form a hybrid energy system. As a result, the wind turbine would produce a higher output during winter, while the solar panels would provide their maximum production during summer. Compared to standalone wind, solar, geothermal,

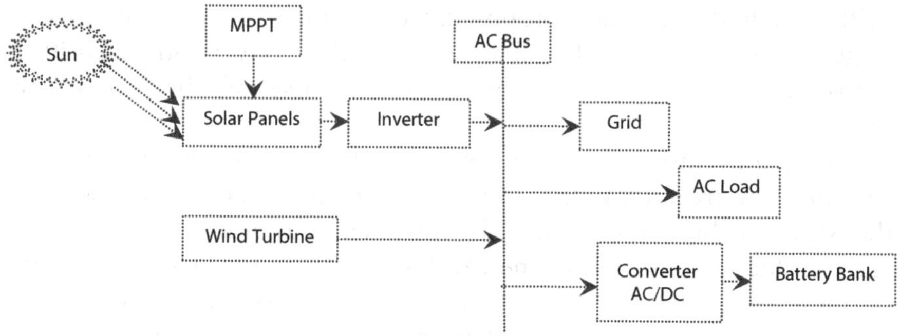

Figure 6.2 Grid-connected HRE system.

or tri-generation energy systems, hybrid energy systems frequently produce superior economic and environmental benefits [22]. The environment (cell temperature and incident radiation of the sun), the inverter's attributes (output, functioning threshold, and working stage), and the grid connection approach, are based on both the stability and dependability of the grid and the characteristics of the energy produced by the inverter. The block diagram of grid connected HRE system is illustrated in Figure 6.2.

6.2.1.2 *Stand-Alone or Off-Grid HRE System*

Stand-alone solutions were utilized in areas without a power grid. A stand-alone system is extremely helpful in remote places without power networks. A standalone device with battery storage generates DC power for a load using solar PV panels [3]. Charge controllers are used to switch off the SPV panel when the battery is fully charged and to turn it back on when the battery charge falls below a certain level when using batteries as a backup system. To be utilized at night and on cloudy days when the sun is not present, batteries must be able to store enough energy throughout the day [23]. The block diagram of off-grid or stand-alone HRE system is illustrated in Figure 6.3.

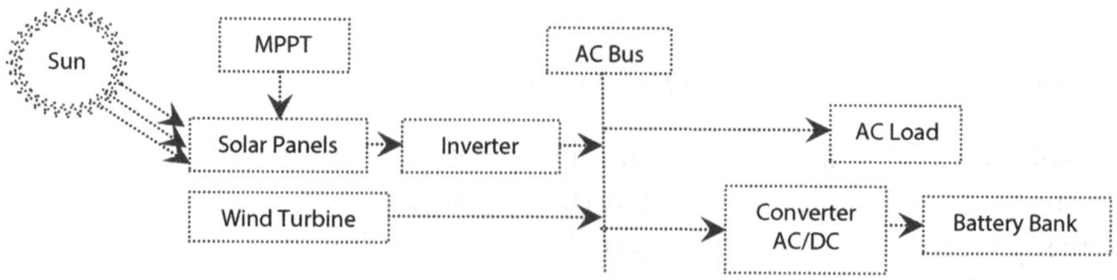

Figure 6.3 Off-grid HRE system.

6.3 PV Array Characteristics

Solar PV systems are composed of individual cells that are essentially semiconducting circuits. Different semiconductor materials can be used to fabricate the cells. Allowing solar photons into the cell causes electrons in the doped silicon to be pushed out and dragged to the lattice of metal conductors, where they travel as direct current and generate electricity [24]. To represent the cell, one- or two-diode equivalent circuits were used. Figure 6.4 shows the single-diode circuit model used in this analysis.

The mathematical formulation for the cell given in Eq. (6.1) is as follows when the parallel resistance and saturation current are modified [15].

$$I_{se} = I_s - I_0 \left\{ exp\left[\frac{q(V + I_{se}R_{se})}{nkT_c} \right] - 1 \right\} - \frac{V + I_{se}R_{se}}{R_{sh}} \tag{6.1}$$

Here, I_o, I_{se}, and V denote the saturation current of the diode, current, and voltage of the solar cell, respectively. R_{sh}, R_{se}, k, q, and T_c are the parallel resistor, series resistor, Boltzmann constant (1.38×10^{-23} J/K), charge of an electron (1.602×10^{-19} C), and temperature of the cell, respectively.

The output power of PV systems varies with the solar radiation and temperature. The output power increases as the current increases, and the sun irradiation increases. However, when the temperature increases, the voltage drops, resulting in a drop in the output power [25]. Figures 6.5 and 6.6 depict the variations in output power, voltage, and current with respect to temperature and solar radiation.

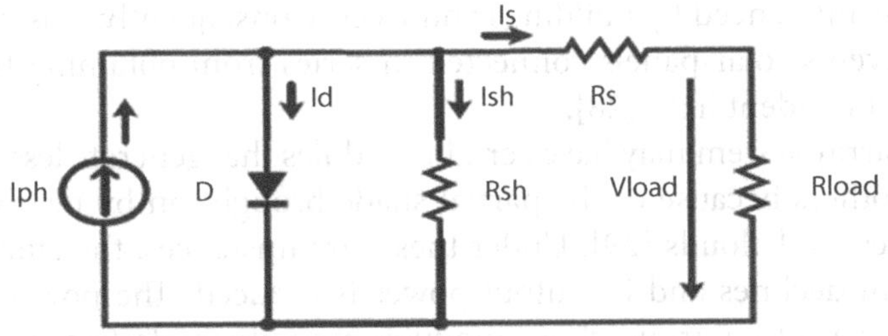

Figure 6.4 Equivalent circuit for a single diode.

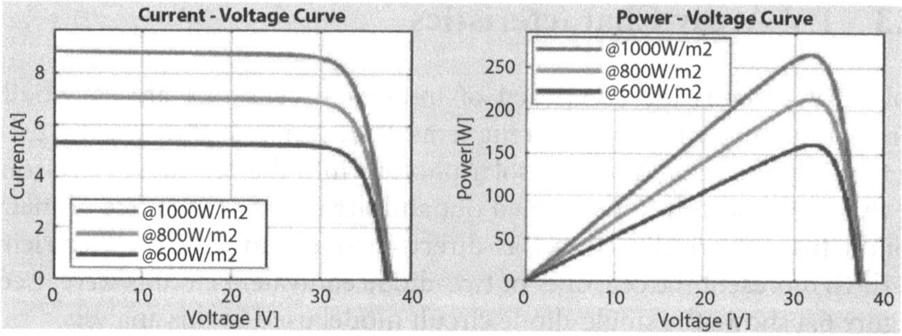

Figure 6.5 Impacts of solar irradiance on the P–V and I–V curves.

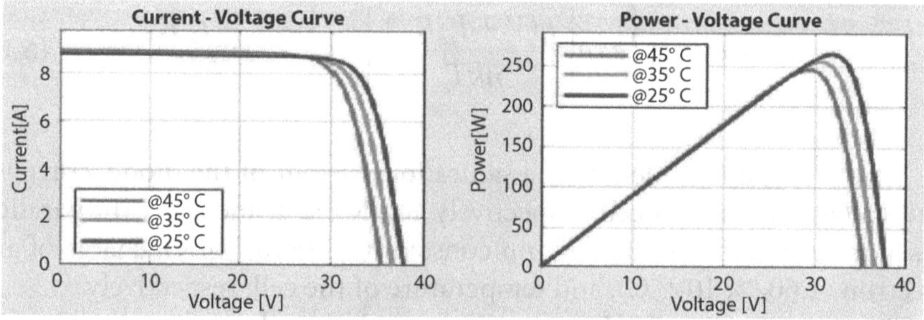

Figure 6.6 Impacts of temperature on the P–V and I–V curves.

6.3.1 The I–V and P–V Curves of a Solar PV Cell Under Partially Shaded Conditions

Partial shadowing is a major risk factor for photovoltaic systems. A shadow is an impression made on a surface (such as a solar panel) by an object that blocks the sun's rays and produces a partial shading effect (PSC) [26]. As a result, constant uniform irradiance is not achievable because of shifting environmental factors, such as rain, clouds, and storms [27]. In addition, shading is influenced by buildings and trees. Consequently, this phenomenon prevents solar panels connected in series from obtaining the same amount of incident light [28].

A PV array system may have certain modules that generate less electricity than others because of the partial shade brought on by nearby structures, trees, and clouds [29]. Under these circumstances, the efficiency of the system declines and its output power is reduced. The power–voltage (P–V) curves of an array whose modules are vulnerable to partial shadows are shown in Figure 6.7 and indicate fluctuations in the output power. When all the system modules are subjected to the same amount of light,

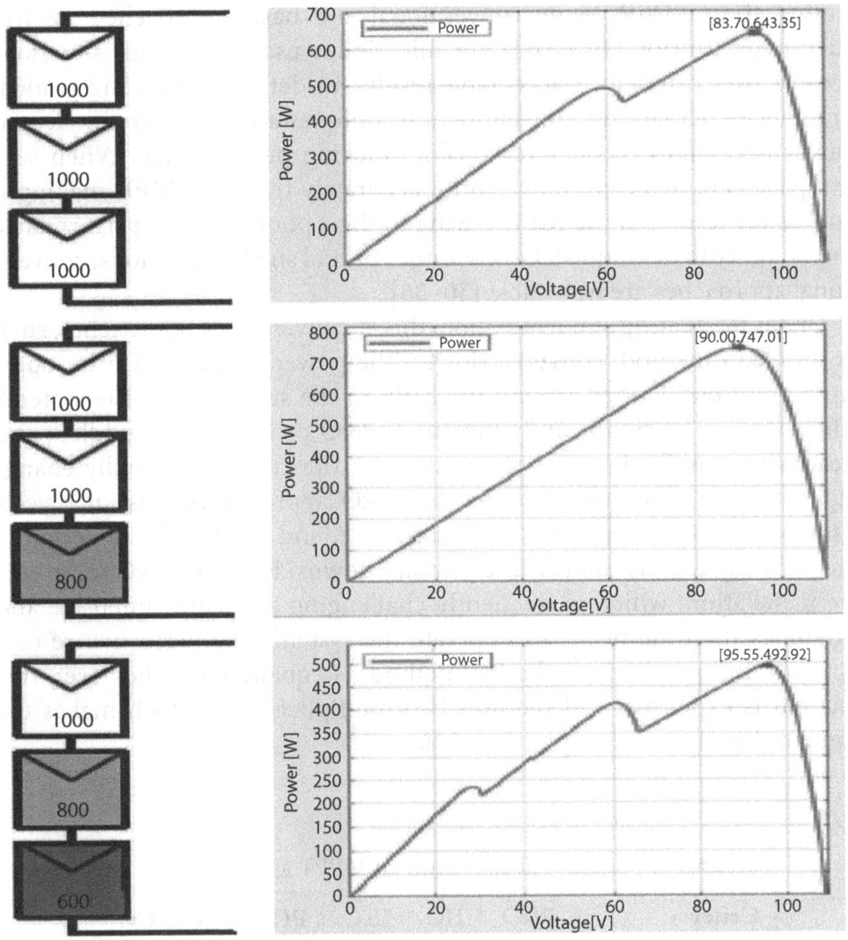

Figure 6.7 The P–V curve's effects in a partly darkened situation.

as shown by the curves, the P–V curve only creates one MPP, known as the global MPP. The largest of these spots is the global MPP, whereas the others are local MPP [30]. However, when the modules were subjected to varying quantities of light, more than one MPP was formed.

6.4 Evaluation of Various MPPT Methods Using Standard Conventional Approaches

One of the key factors for selecting the best MPPT approach is identifying the one that has a small tracking error, rapid effectiveness, and fluctuation around the GMMP. Many conventional and basic approaches are frequently used in practice; they are suitable for use in everyday situations because they can deliver acceptable results. Under clear skies and gradual irradiance fluctuations, conventional techniques are generally simple and inexpensive and can offer reasonable tracking performance. When several peaks are detected, they cannot accurately track the MPP, making it impossible to obtain the MPP. Owing to their poor tracking performance and inability to distinguish GMPP under partial shade conditions, conventional approaches are not viable [30–35].

Under fluctuating sun irradiation, direct conventional approaches, such as the P&O method, converge slowly. The power loss results from operating point oscillations around the MPP in the steady state. The system's efficiency can be significantly impaired, P&O can become muddled, and steady-state oscillations can appear. When the climate is rapidly changing, the INC approach surpasses the P&O approach because it reveals less uncertainty, has greater noise tolerance, and lessens the uncertainty caused by system dynamics. The primary downsides of the INC technique are installation, which is frequently challenging and labor-intensive and lengthens the time needed to operate the system. The performance of a direct technique, such as the PC method, is equally good; however, this method is expensive and requires two multiplications, which makes the control circuit more complex [34, 35].

Table 6.1 Evaluation of conventional MPPT approaches.

Criteria	**P&O**	**HC**	**INC**	**PC**	**CV**	**CC**
Complexity	L	L	M	M	L	L
Robustness	H	H	H	H	L	L
Efficiency	M	M	RH	RH	L	L
Sensors	3	3	3	3	1	2
Periodic tuning	N	N	N	Y	Y	Y
Tracking speed	S	S	M	M	F	F
Cost	M	M	M	M	L	L
Type	ON	ON	ON	OF	OF	OF

*__Note:__ L, Low; M, Medium; H, High; RH, Relatively High; N, No; Y, Yes; S, Slow; F, Fast; ON, Online; OF, Offline; 1, I; 2, V; and 3, I & V.

Indirect conventional techniques, such as CV and CC, are affordable. When scanning the complete control range using the CV technique, system functioning is interrupted, resulting in power losses. As a result, the actual power consumed is not seen as coming from the panels. In other words, because it is assumed that the maximum point is computed for a certain O.C. voltage, the harvested power will not be the maximum if the operating point is inaccurate or slightly off. Knowing the precise value of the short-circuit current proportionality factor affects the accuracy of the CC technique [36]. For each PV module used in each installation, along with the matching value at the MPP, and any variations caused by irradiance and aging. The conventional approaches are compared in Table 6.1 [2, 36–38]. Table 6.1 displays an aggregate overview of the considered s. Only conventional approaches are used for the bulk of comparisons to help identify the key assessment criteria.

6.5 Evaluation of Various MPPT Methods Using Advanced Approaches (Metaheuristic Optimization Approaches)

Numerous variables have been suggested for the assessment of MPPT. These characteristics are usually used to provide a general judgement; some of them are even approximations because they cannot be measured precisely. This section compares MPPT strategies utilizing advanced methodologies (metaheuristic optimization techniques) from various perspectives, as indicated in Table 6.2 [2, 39–46]. A typical summary of the nominated soft computing studies is presented in Table 6.2.

It uses soft computing approaches for the bulk of comparisons to highlight the differences depending on various parameters. It should be emphasized that several testing situations, such as experimental and simulation tests, were used to verify the findings in Tables 6.1 and 6.2. Some of the studies that have persisted are based on simulations, whereas others are based on experiments. As previously stated, it is an estimate based on most of the outcomes.

The MPPT techniques are not ranked in a particular order. The ranking criteria include applicability, tracking speed, complexity, and cost. As a conclusion from Tables 6.1 and 6.2, Figure 6.8 may give a summary of MPPT ranking depending on the above parameters [2, 36–46]. The major goal of this ranking is to make it easier to choose approaches using standard well-defined parameters.

Table 6.2 Evolution of various metaheuristic optimization-based MPPT approaches.

Criteria	FLC	ANN	PSO	GA	ABC	CA	BA	CSO	ACO	JA	DE	GWO	FA
Type	1	1	2	2	2	2	2	2	2	2	2	2	2
Sensors	4	3	3	3	3	3	3	3	3	3	3	3	3
Oscillation around MPP	N	N	N	N	N	N	N	N	N	N	N	N	N
Periodic tuning	Y	Y	N	N	N	N	N	N	N	N	N	N	N
PV array dependency	Y	Y	N	N	N	N	N	N	N	N	N	N	N
Complexity	H	H	H	H	H	H	H	H	H	H	H	H	H
Tracking speed	A	A	F	F	F	F	F	VF	F	VF	M	V	V
Cost	H	H	H	H	H	H	H	H	H	H	H	H	H
Efficiency	L	L	H	H	H	H	H	H	H	H	H	H	H

***Note:** N, No; Y, Yes; H, High; A, Average; F, Fast; M, Medium; VF, Very Fast; L, Low; 1, AI; 2, BI; 3, V,I; and 4, V,I,G.

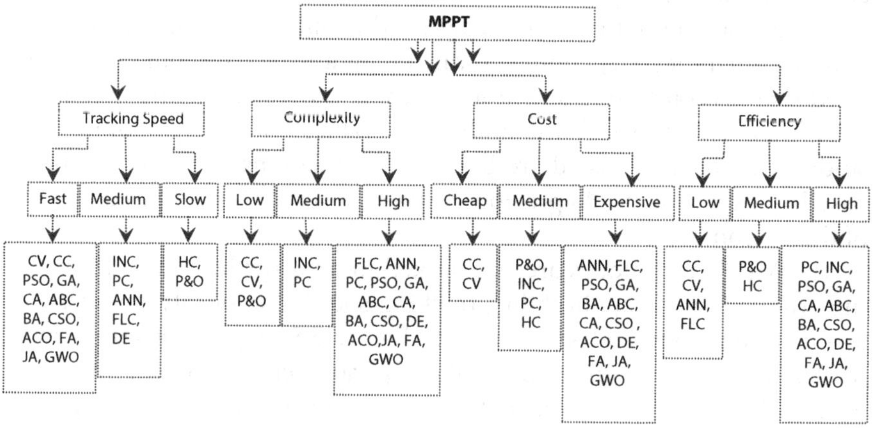

Figure 6.8 Overview of MPPT ranking factors.

6.5.1 Benefits and Restrictions of MPPT Approaches Based on Metaheuristic Optimization

Soft computing approaches provide solutions and efficient adaptable tracking strategies that can deal with changing weather and scenarios with PS, even if they are the most difficult to use, and demonstrate superior tracking performance compared to conventional techniques [37–42]. When approaching the MPP, which outperforms conventional approaches, the velocity drops dramatically or becomes almost zero.

AI techniques perform well under typical situations. The reviewed studies have shown that the original AI approaches are unable to precisely monitor GMMPs under PSCs. To obtain reliable results, AI-based approaches, such as ANN, must first undergo extensive adjustments to obtain the best MPP. They must also undergo thorough training by utilizing many measurements before being used in the MPPT control unit. The development of a rule base and membership function is essential for FLC controllers. There is no accepted technique for accurately defining the controller parameters. When the operational conditions change significantly and expert advice provided is inaccurate, the FLC with associated parameters is unsuitable. Utilizing an adaptive fuzzy logic controller (AFLC) can solve this issue. To obtain the best performance, the AFLC parameters may be altered [43–47].

In comparison to analytical procedures, BI-based solutions also have the problem with sluggish tracking speeds. The primary factor in the original MPPT is the initialization value. Therefore, it is important to carefully choose the initialization settings. By reducing the number of acceleration factors in the velocity expression, the optimization is simplified, similar to the modified PSO, which reduces the velocity factor in accordance with the distance between two peaks. According to the FA approach, each firefly's location is gradually shifted towards the direction of fireflies that are more brilliant. The modified firefly MFA algorithm has been suggested as a solution to the problems associated with the conventional FA technique.

Hybrid approaches, which are technically sound, are the best way to tackle problems using unique methods. By employing various operating concepts, the MPP can be tracked and its performance improved. The complicated control structure and high cost of hardware implementation are the disadvantages of this strategy. If the GMPP is placed beyond the search zone, a combination of conventional and advanced approaches cannot guarantee convergence. Complex calculations are required for advanced hybrid techniques and many parameters must be chosen for optimization.

6.6 Conclusion and Future Scope

This chapter offers a comprehensive review of various MPPT optimization techniques, which may be classified into different groups: conventional, meta-heuristic, and AI. A thorough explanation of the operational measurements is presented in this chapter. Subsequently, based on the publications that have been studied, several MPPT techniques, including those that are more recent, are evaluated while considering certain aspects. The following factors were acknowledged: algorithm difficulty, tracking speed, expenses, sensory parameters, monitoring accuracy, SPV array dependency, steady-state fluctuation, and A/D. The benefits and drawbacks of MPPT methods are also tabulated. In addition, this study argues that the best method to address PSC issues is to use MPP controllers. As a result, this has opened the door for extensive studies. Under uniform irradiance conditions, the established approaches are effective and have simple algorithms. However, compared with more advanced algorithms, the tracking rates of classical algorithms are sluggish. Furthermore, as these methods produce the best outcomes in PSCs, intelligent strategies are more prominent. Additionally, intelligent algorithms simplify data storage and are not dependent on mathematical calculations. In contrast, bio-inspired meta-heuristic methods can be used with any solar PV-based HRE system. The specifications of the PV panel are not necessary for these solutions. In summary, an analyst or researcher may use this review chapter as a useful resource when choosing an MPPT technique to achieve a certain goal. In addition, a parallel comparison of MPPT approaches is presented in this chapter. This report is the result of a thorough literature assessment and may be used as a resource and tool for scholars studying the PV system sector.

The expense of the algorithms, difficulty of hardware implementation, measurement inaccuracy, and promotion of current sensorless MPPT may all be summed up as issues facing the currently used MPPT methods. Finding an MPP tracker that is easier to use, quicker, faster, cheaper, and more reliable is recommended. More studies are needed for this assignment.

Acknowledgment: The MCN "IU/R&D/2022-MCN0001674," which was provided by Integral University, Lucknow, is acknowledged by the authors. Writers alone bear full responsibility for the assertions stated in this chapter.

References

1. Boghdady TA, Kotb YE, Aljumah A, Sayed MM. Comparative Study of Optimal PV Array Configurations and MPPT under Partial Shading with Fast Dynamical Change of Hybrid Load. *Sustainability*, 2022; 14(5):2937. https://doi.org/10.3390/su14052937.
2. A. O. Baba, G. Liu , and X. Chen, "Classification and Evaluation Review of Maximum Power Point Tracking Methods", *Sustain. Futures*, 2 (2020) 100020. https://doi.org/10.1016/j.sftr.2020.100020.
3. Khan, A.A.; Minai, A.F.; Pachauri, R.K.; Malik, H. Optimal Sizing, Control and Management Strategies for Hybrid Renewable Energy Systems: A Comprehensive Review. *Energies*, 2022, 15(17), 6249. https://doi.org/10.3390/en15176249
4. El-Sattar HA, Kamel S, Sultan H, Tostado-Véliz M, Eltamaly AM, Jurado F. Performance Analysis of a Stand-Alone PV/WT/Biomass/Bat System in Alrashda Village in Egypt. *Appl. Sci.*, 2021; 11(21):10191. https://doi.org/10.3390/app112110191
5. M. Kumar, A. F. Minai, A. A. Khan and S. Kumar, "IoT based Energy Management System for Smart Grid," *2020 International Conference on Advances in Computing, Communication & Materials (ICACCM)*, 2020, pp. 121-125, doi: 10.1109/ICACCM50413.2020.9213061.
6. Blaabjerg F, Ionel DM. *Renewable Energy Devices and Systems with Simulations in MATLAB and ANSYS®*. CRC Press 2017. p. 67-90.
7. A Ibrahim, A Fudholi, K Sopian, MY Othman, MH Ruslan, Efficiencies and improvement potential of building integrated photovoltaic thermal (BIPVT) system. *Energy Convers. Manage.*, 77 (2014) 527–534.
8. K Masuko, M Shigematsu, T Hashiguchi, D Fujishim , M Kai, N Yoshimura, T Yamaguchi, Y Ichihashi, T Mishima, N Matsubara, T Yamanishi, Achievement of more than 25% conversion efficiency with crystalline silicon heterojunction solar cell. *IEEE J. Photovoltaics*, 4 (2014) 1433–1435.
9. J Marcos, L Marroyo, E Lorenzo, D Alvira, E Izco, Power output fluctuations in large scale PV plants: one year observations with one second resolution and a derived analytic model. *Prog. Photovoltaics Res. Appl.*, 19 (2011) 218–227.
10. F Belhachat, C. Larbes, A review of global maximum power point tracking techniques of photovoltaic system under partial shading conditions. *Renew. Sust. Energ. Rev.*, 92 (2018) 513–553.
11. R Rajesh, MC. Mabel, A comprehensive review of photovoltaic systems. *Renew. Sust. Energ. Rev.*, 51 (2015) 231–248.
12. Minai, A.F., Usmani, T., Uz Zaman, S., Minai, A.K. (2022). Intelligent Tools and Techniques for Data Analytics of SPV Systems: An Experimental Case Study. In: Malik, H., Ahmad, M.W., Kothari, D. (eds) *Intelligent Data Analytics for Power and Energy Systems*. Lecture Notes in Electrical Engineering, vol 802. Springer, Singapore. https://doi.org/10.1007/978-981-16-6081-8_16

13. Y Wang, Y Li, X Ruan, High-accuracy and fast-speed MPPT methods for PV string under partially shaded conditions. *IEEE Trans. Ind. Electron.*, 63 (2015) 235–245.
14. H Rezk, A Fathy, AY Abdelaziz, A comparison of different global MPPT techniques based on meta-heuristic algorithms for photovoltaic system subjected to partial shading conditions. *Renew. Sust. Energ. Rev.*, 74 (2017) 377–386.
15. A. A. Khan, A. F. Minai, L. Devi, Q. Alam and R. K. Pachauri, "Energy Demand Modelling and ANN Based Forecasting using MATLAB/Simulink," *2021 International Conference on Control, Automation, Power and Signal Processing (CAPS)*, 2021, pp. 1-6, doi: 10.1109/CAPS52117.2021.9730746.
16. B Bendib, H Belmili, F Krim, A survey of the most used MPPT methods: Conventional and advanced algorithms applied for photovoltaic systems. *Renew. Sust. Energ. Rev.*, 45 (2015) 637–648.
17. JP Ram, N. Rajasekar, A new global maximum power point tracking technique for solar photovoltaic (PV) system under partial shading conditions(PSC). *Energy*, 118 (2017) 512–525.
18. MA Elgendy, B Zahawi, DJ Atkinson, Assessment of the incremental conductance maximum power point tracking algorithm. *IEEE Trans. Sustain. Energy*, 4 (2012) 108–117.
19. J Ahmed, Z. Salam, A critical evaluation on maximum power point tracking methods for partial shading in PV systems. *Renew. Sust. Energ. Rev.*, 47 (2015) 933–953.
20. Çelik O, Teke A., A Hybrid MPPT method for grid connected photovoltaic systems under rapidly changing atmospheric conditions. *Electr. Power Syst. Res.*, 152 (2017) 194–210.
21. MS mahmoudian, R Rahmani, S Mekhilef, AM Oo, A Stojcevski, TK Soon, AS Ghandhari, Simulation and hardware implementation of new maximum power point tracking technique for partially shaded PV system using hybrid DEPSO method. *IEEE Trans. Sustain. Energy*, 6 (2015) 850–862.
22. Fatima, K., Minai, A.F., Malik, H. (2022). Intelligent Approach-Based Maximum Power Point Tracking for Renewable Energy System: A Review. In: Malik, H., Ahmad, M.W., Kothari, D. (eds) *Intelligent Data Analytics for Power and Energy Systems*. Lecture Notes in Electrical Engineering, vol 802. Springer, Singapore. https://doi.org/10.1007/978-981-16-6081-8_19
23. M. R. Shaikh, S. Shaikh, S. Waghmare, S. Labade, and A. Tekale, "A review paper on electricity generation from solar energy," *Int. J. Res. Appl. Sci. Eng. Technol.*, vol. 5, no. 9, pp. 1884-1889,2017.
24. K Ishaque, Z. Salam, An improved modeling method to determine the model parameters of photovoltaic (PV) modules using differential evolution (DE). *Sol. Energy*, 85 (2011) 2349–2359.
25. R. K. Pachauri, A. K. Gupta, A. F. Minai, M. Kumar and Shashikant, "Su-Do-Ku Game Puzzle for Improving Shade dispersion factor on PV Array Systems under PSCs: Experimental Validation," *2022 International*

Conference on Emerging Smart Computing and Informatics (ESCI), 2022, pp. 1-6, doi: 10.1109/ESCI53509.2022.9758362.

26. HA Sher, AF Murtaza, A Noman, KE Addoweesh, K Al-Haddad, M Chiaberge, A new sensorless hybrid MPPT algorithm based on fractional short-circuit current measurement and P&O MPPT. *IEEE Trans. Sustain. Energy*, 6 (2015) 1426–1434.
27. G Yüksek, AN. Mete, A hybrid variable step size MPPT method based on P&O and INC methods, in: *2017 10th International Conference on Electrical and Electronics Engineering (ELECO)*, IEEE, 2017, pp. 949–953.
28. LL Jiang, DR Nayanasiri, DL Maskell, DM Vilathgamuwa, A hybrid maximum power point tracking for partially shaded photovoltaic systems in the tropics. *Renew. Energy*, 76 (2015) 53–65.
29. S. O. Gyan Torto, R. Kumar Pachauri, A. F. Minai, A. Siddiqui and A. Sharma, "PV Module Reconfigurable Approach using SM for Power Loss Reduction under Detrimental Shadowing Conditions," *2022 3rd International Conference for Emerging Technology (INCET)*, 2022, pp. 1-6, doi: 10.1109/INCET54531.2022.9823969.
30. R. K. Pachauri, S. Motahhir, A. K. Gupta, M. Sharma, A. F. Minai, M. S. Hossain, A. Yassine, Game theory based strategy to reconfigure PV module arrangements for achieving higher GMPP under PSCs: Experimental feasibility. *Energy Rep.*, Volume 8, 2022, Pages 10088-10112, ISSN 2352-4847, https://doi.org/10.1016/j.egyr.2022.08.006.
31. SZ Said, L. Thiaw, "Performance of Artificial Neural Network and P Swarm Optimization Technique based Maximum Power Point Tracking for Photovoltaic System Under Different Environmental Conditions", In *J. Phys. Conf. Ser.*, IOP Publishing 1049 (2018) 012047.
32. M. Naseem, M. A. Husain, J. D. Kumar, M. W. Ahmad, A. F. Minai and A. A. Khan, "P Swarm Optimization based Maximum Power Point Tracking Technique for Solar PV System under Partially Shaded conditions," *2021 International Conference on Control, Automation, Power and Signal Processing (CAPS)*, 2021, pp. 1-6, doi: 10.1109/CAPS52117.2021.9730703.
33. Minai, A F, Husain, M A, Naseem, M and Khan, A. A. "Electricity demand modeling techniques for hybrid solar PV system" *Int. J. Emerging Electr. Power Syst.*, vol. 22, no. 5, 2021, pp. 607-615. https://doi.org/10.1515/ijeeps-2021-0085
34. N. Bilgin, and İ. Yazici, "Comparison of Maximum Power Point Tracking Methods Using Metaheuristic Optimization Algorithms for Photovoltaic Systems" *Sakarya University J. Sci.*, 25(4), Turkey. 1075-1085, 2021.
35. R. K. Pachauri, A. F. Minai, V. Pandey, Shashikant, V. Goyal and R. Swain, "Optimal Placement of Modules in PV Array Systems to Achieve Higher GMPP under Obscured Irradiations," *2021 International Conference on Control, Automation, Power and Signal Processing (CAPS)*, 2021, pp. 1-7, doi: 10.1109/CAPS52117.2021.9730692.

36. Naseem, M., Husain, M.A., Minai, A.F. *et al.* Assessment of Meta-Heuristic and Classical Methods for GMPPT of PV System. *Trans. Electr. Electron. Mater.*, 22, 217–234 (2021). https://doi.org/10.1007/s42341-021-00306-3.
37. Minai, A.F., Malik, H. (2021). Metaheuristics Paradigms for Renewable Energy Systems: Advances in Optimization Algorithms. In: Malik, H., Iqbal, A., Joshi, P., Agrawal, S., Bakhsh, F.I. (eds) *Metaheuristic and Evolutionary Computation: Algorithms and Applications*. Studies in Computational Intelligence, vol 916. Springer, Singapore. https://doi.org/10.1007/978-981-15-7571-6_2
38. M.A. Husain *et al.* Fast and precise global maximum power point tracking techniques for photovoltaic system." *IET Renewable Power Gener.*, 13.14 (2019): 2569-2579. https://doi.org/10.1049/iet-rpg.2019.0244.
39. M. A. Siddiqui, M. Nishat Anwar, A. F. Minai, A. Ahmad Khan, M. Naseem and A. Jabbar, "A Direct Synthesis based Sliding Mode Control of a Nonlinear Continuous Stirred Tank Reactor," *IECON 2022 – 48th Annual Conference of the IEEE Industrial Electronics Society*, Brussels, Belgium, 2022, pp. 1-6, doi: 10.1109/IECON49645.2022.9969082.
40. A Gautam, A J Ansari, and A. A. Khan "Review of Fuzzy Logic Applications in Performance Enhancement of Solar Based Power System " in *International Journal of Research and Development in Applied Science and Engineering (IJRDASE)*, Vol. 8, Issue 2 (December 2015) (ISSN 2454 –6844).
41. Husain, M.A., Jain, A., & Tariq, A. (2016). A novel fast mutable duty (FMD) MPPT technique for solar PV system with reduced searching area. *J. Renew. Sustain. Energy*, 8, 054703.https://doi.org/10.1063/1.4963314.
42. Khan, A.A., Minai, A. F., "Introduction to Grid-Forming Inverters", B. (Eds.), (2023) *Grid-Forming Power Inverters: Control and Applications* (1st ed.). page 1-14, ISBN-9781003302520, CRC Press. https://doi.org/10.1201/9781003302520-1.
43. Naseem, M., Husain, M. A., Minai, A. F., Khan, A. N., Amir, M., Dinesh Kumar, J., & Iqbal, A. (2021). Assessment of meta-heuristic and classical methods for GMPPT of PV system. *Trans. Electr. Electron. Mater.*, 22, 217-234. https://doi.org/10.1007/s42341-021-00306-3.
44. M.A. Husain *et al.* Comparative assessment of maximum power point tracking procedures for photovoltaic systems, *Green Energy Environ.*, 2 (2017) 5-17. https://doi.org/10.1016/j.gee.2016.11.001
45. Faiz Minai, A.; Khan, A.A.; Pachauri, R.K.; Malik, H.; García Márquez, F.P.; Arcos Jiménez, A. Performance Evaluation of Solar PV-Based Z-Source Cascaded Multilevel Inverter with Optimized Switching Scheme. *Electronics*, 2022, 11, 3706. https://doi.org/10.3390/electronics11223706.
46. M. Naseem *et al.*, "A Spider Monkey Optimization Based Global Maximum Power Point Tracking Technique for Photovoltaic Systems," *2022 2nd International Conference on Emerging Frontiers in Electrical and Electronic*

Technologies (ICEFEET), Patna, India, 2022, pp. 1-6, doi: 10.1109/ICEFEET51821.2022.9847903.

47. A.A. Khan, A. F. Minai (2023)"A Strategic Review: The Role of Commercially Available Tools for Planning, Modeling, Optimization, and Performance Measurement of Photovoltaic Systems", *Energy Harvesting Syst., (EHS)*, ISSN: 2329-8766. https://doi.org/10.1515/EHS-2022-0157.

7

Intelligent Modeling and Estimation of Solar Radiation Data Using Artificial Intelligence

Ahmad Neyaz Khan[1]*, Sarosh Patel[2], Asif Khan[3], Asad Malik[4], Mohd Fazil[5] and Saqib Qamar[6]

[1]*Department of Computer Science and Application, KL University, Vijayavada, India*
[2]*School of Engineering, University of Bridgeport, Bridgeport, CT, USA*
[3]*Department of Computer Application, Integral University, Lucknow, Uttar Pradesh, India*
[4]*School of Information Technology, Monash University, Subang Jaya, Selangor, Malaysia*
[5]*Centre for Transformative Learning, University of Limerick, Limerick, Ireland*
[6]*Department of Physics, Umea University, Linnaeus vag 20, Umea, Sweden*

Abstract

Artificial Intelligence (AI) is currently being employed in a variety of fields, including engineering, economics, medicine, military, and marine. Complex systems have also been intelligently modeled, identified, optimized, predicted, forecasted, and controlled using AI. Our goal is to provide an overview of intelligent simulation models for solar power and photovoltaic (PV) forecasting. The works considered in this chapter demonstrate the benefits of employing AI-based photovoltaic data forecasting in distant places where almost no equipment exists to monitor this information. Moreover, they demonstrated the potential of AI as a design tool for solar radiation data prediction and forecasting, particularly parameters related to photovoltaic (PV) systems. The performance and sizing are heavily influenced by solar radiation. This chapter provides an overview of the current and promising uses of artificial intelligence in the area of photovoltaic systems. It covers a wide variety of issues and provides a complete overview of the field.

Keywords: Artificial intelligence, sustainable energy, solar PV, forecasting, intelligent modeling

**Corresponding author*: ahmadnk500@gmail.com

Mohammed Aslam Husain, Md Waseem Ahmad, Farhad Ilahi Bakhsh, P. Sanjeevikumar and Hasmat Malik (eds.) Photovoltaic Systems Technology, (123–150) © 2024 Scrivener Publishing LLC

7.1 Introduction

The use of alternative [1–4] or sustainable energy has acquired widespread support in response to acute energy issues and environmental damage caused by conventional fuels. Solar energy is among the fastest-growing alternative energy sources. Solar radiation, also known as solar insolation, is the amount of energy received from the sun over a specific surface area. Accurate estimation of solar radiation data is essential for various applications such as solar energy forecasting, photovoltaic system design, and agriculture. However, traditional methods [5] of solar radiation estimation are often based on statistical models, which can be unreliable owing to the complex and nonlinear nature of solar radiation data.

The swift expansion of PV faces obstacles stemming from its reliance on weather conditions, causing an irregular energy supply from PV modules. This creates difficulties in maintaining market stability, planning the reserve capacity, and conducting electricity market bidding. The integration of large-scale PV systems presents a significant hurdle for electric system operators, as it impedes grid management. Poor management can result in an unstable voltage and heightened grid instability. Power production from the sun is affected by a wide variety of environmental conditions, such as solar irradiance, cloud cover, heat, moisture, barometric pressure, and air velocity. It is difficult to reliably and accurately anticipate [6, 7] power output owing to unpredictable atmospheric phenomena, which might drastically alter any moment.

The field of artificial intelligence (AI) is still developing, where changes occur rapidly at first and then gradually across periods. The scientific world has shifted its emphasis towards new subjects and uses. AI's capacity to resolve issues and come up with new ideas may grow as technology advances. Scientists have expanded on prior work, put it into practice today, and imagine opportunities ahead. AI is important [8–10] for modeling and estimation of power in solar PV systems because it can provide accurate [11, 12] and efficient predictions [13] of energy generation, which is critical for optimizing system performance and reducing costs.

One of the key challenges [14] in solar PV systems is accurately predicting [15] the amount of energy that will be generated by the system, given factors [16] such as weather conditions, solar panel efficiency, and shading. Traditional models [17] for estimating solar PV system outputs are often based on simplified assumptions and do not consider the complex interactions between these factors. This can lead to significant errors in energy production estimates and impact the overall performance of the system.

On the other hand, AI-based models can learn from historical data and use sophisticated algorithms to predict energy output more accurately. For example, machine-learning algorithms can analyze data from multiple sensors and weather forecasts to generate real-time predictions of energy output, which can be used to adjust system operations in real time to maximize efficiency and performance.

Moreover, AI can be used for fault detection [18–20] and diagnostics [21–23] in solar PV systems. By analyzing data from sensors and other sources, AI algorithms can detect anomalies in the system performance and identify the root cause of the problem. This can help operators to quickly identify and fix issues, reduce downtime and increase the overall reliability of the system. Thus, AI is important for modeling and estimating the power in solar PV systems because it can provide accurate and efficient predictions of energy generation, enabling operators to optimize system performance, reduce costs, and improve overall reliability.

Neural networks, support vector machines, machine learning (ML), knowledge-based systems, fuzzy logic, and heuristic algorithms are some of the most frequently utilized AI techniques that can learn complex patterns in solar radiation data and make accurate predictions. Artificial intelligence (AI) can be used to model and estimate the solar radiation in several ways:

1. To develop models that accurately predict solar radiation based on past data and other relevant factors such as weather conditions, cloud cover, and geographic location.
2. Real-time estimations of solar radiation can be obtained by analyzing satellite photos and other data sources, which can be utilized to improve the generation of energy in solar power facilities.
3. Identifying patterns in solar radiation data and detecting anomalies, which can help identify potential issues in the solar energy system and improve overall performance.

It is impossible to overlook the potential of artificial intelligence (AI) approaches to dramatically increase the precision and efficacy of solar irradiation simulations and predictions, making it an effective tool in the area of sustainable energy.

In this chapter, we propose a comprehensive assessment of the status of solar irradiance prediction and modeling using AI approaches. Here, artificial intelligence methods are briefly introduced before their use in modeling and predicting solar radiation. The overall organization of this chapter is *Section 7.2*: The Solar-AI Span: Background and Literature Review, which has been introduction to the solar-AI span, historical background with an overview of the use of AI approaches in solar irradiance modeling and prediction along with a review of relevant literature on the topic. *Section 7.3* follows the modeling and prediction of data on solar irradiance using AI approaches. This section explains the AI approaches used for modeling and predicting of solar irradiance methodologies in this chapter and the evaluation of the accuracy of the models developed using the different AI approaches. *Section 7.4* deals a detailed comparative analysis of different AI approaches used in modeling and prediction of data on solar radiation, and a comparative analysis of the performance of the different AI approaches used is discussed along with their strengths and limitations, and identification of the most effective AI approach for modeling and prediction of solar irradiance. *Section 7.5* contains an interpretation of the results obtained from the study and a discussion of the implications of the study's findings for the field of solar energy identification of areas for further research. *Section 7.6* has the summary of the chapter's main findings, concludes the implications for the use of AI approaches in solar irradiance modeling and prediction and provides suggestions for future research.

7.2 The Solar-AI Span: Background and Literature Review

The application of AI techniques in the modeling and estimation of solar radiation in PV systems has gained significant attention in recent years. Because the profitability [24] and reliable functioning of this system

Figure 7.1 Shuffled frog leaping algorithm (SFLA) jumping method.

are threatened by the instability [18–23] that solar power production is weather-dependent, researchers have been working towards developing accurate and efficient models to estimate solar radiation using artificial intelligence algorithms.

Step 1: Initialization Choose the number of memeplexes, M, and the number of frogs in each memeplex, N. The total population size, F, is calculated as $F = M * N$.

Step 2: Virtual Population Creation From the available space, randomly sample F virtual frogs denoted as $U(1), U(2), \ldots, U(F)$. Calculate the competency value of each U(i) to assess their suitability.

$U(i) = (Ui1, Ui2, \ldots, Uid)$, where d represents the number of decision variables.

Step 3: Sorting and Ranking the Frogs Arrange the frogs in a descending order based on their competency values and store them in the array $X = \{U(i), f(i) \,\&\, i = 1\ldots, F\}$. Determine the position of the best frog, Px, in the population ($U = Px(1)$).

Step 4: Dividing Frogs into Memeplexes Divide the array X into Memeplexes, Y, each containing N frogs.

Step 5: Evolution of Memeplexes Apply the local search frog jump algorithm to each memeplex, Yk ($k = 1,2,3,\ldots, M$), to evolve them.

Step 6: Combining Memeplexes After a certain number of evolutions have been completed in each memeplex, combine the memeplexes ($Y1, \ldots, Ym$) in array X such that $X = \{Y(k), k = 1,2,\ldots, M\}$. Update the position of the best frog in the population, Px.

Step 7: Convergence Study Check if the convergence criteria have been met. If yes, terminate the algorithm. If no, return to step 4 for global search.

Artificial Neural Networks (ANNs) are used to estimate the solar radiation [25–27]. Several studies [25–28] have proved that ANNs provide more accurate predictions than traditional statistical methods.

Another popular AI technique for solar radiation estimation is the support vector machines (SVMs). SVM models have modeled complex solar radiation-environmental factors, such as wetness, temperature, and atmospheric pressure. One study [28] concluded that ANNs and SVMs perform the best, particularly owing to their capacity to quickly adjust to different environmental circumstances. Fuzzy logic models [29–31] also incorporate uncertainty in the input variables, making them well suited for modeling real-world problems. Several studies have shown that fuzzy models provide more accurate predictions than traditional statistical methods [5]. In the past few years, deep learning (DL) algorithms [32–38] for estimating solar radiation have become more popular. Spatial trends in sun radiation data are well captured by convolutional neural networks (CNNs). The solar radiation time trends were modeled using RNNs.

A. *Meta-heuristic* algorithms [39, 40], such as shuffled frog leaping algorithm [41] in Figure 7.1, ant colony optimization [42], and multiverse

1. Initialize a matrix of pheromones and a matrix of heuristic information.
2. Repeat the following for a specified number of iterations:
    a. For each ant:
        i. Choose a random starting city.
        ii. While there are unvisited cities:
            1. Compute the probability of moving to each unvisited city based on the pheromones and heuristic information.
            2. Choose the next city to visit based on the probabilities.
            3. Move to the chosen city and update the ant's trail.
        iii. Update the pheromone matrix based on the ant's trail.
    b. Update the pheromone matrix globally based on the best trail found so far.
3. Return the best trail found.

Figure 7.2 A simple pseudocode for applying ACO algorithm to optimize solar radiation data forecasting.

optimization [43, 44], are essential tools for optimizing photovoltaic (PV) power forecasting problems. Figure 7.1 gives the steps are a summary of the two primary phases of global search and local search in the meta-exploration approach of the SFLA algorithm. Figure 7.2 gives the pseudocode for the ACO algorithm. Here, the pheromone matrix contains information about the paths taken by the ants, whereas the heuristic information matrix contains information about the distances between cities. The algorithm works by simulating the behavior of ants, which choose paths based on both pheromone and distance information and updates the pheromone matrix based on the quality of the paths they find. The algorithm continues to iterate until a specified number of iterations are reached, at which point it returns the best path. These algorithms offer an effective way to optimize complex forecasting models by searching through a vast search space for the best solutions. Among the most popular optimization algorithms used for PV power forecasting are the genetic algorithm (GA) [41, 45, 46] and particle swarm optimization [8, 47–49]. These algorithms are inspired by natural phenomena and have been shown to significantly improve forecasting performance.

For example, these algorithms have been used to reduce the normalized mean absolute error (NMAE) of various forecasting models. Figure 7.3 shows the pseudocode for the application of GA for Solar PV forecasting is shown, where a candidate solution represents a set of parameters for a model used to forecast solar radiation data. The population is a collection of such candidate solutions, and each individual in the population has a fitness score that measures how well it performs in forecasting data. The Genetic Algorithm works by selecting parents from the fittest individuals, creating offspring by combining their genetic information through crossover and mutation, and evaluating the fitness of the offspring

```
1. Create a population of potential solutions from scratch.
2. Repetition for a predetermined quantity of generations:
      a. Assess each potential solution's fitness based on how
   well it predictions solar radiation data.
      b. Reproduce with fit parents.
      c. Combine the genetic makeup of the parents to produce
   kids using crossover and mutation.
      d. Add offspring to the population based on fitness.
      e. Choose the fittest individual to make up the population's
   future.
3. Return the best candidate solution found.
```

Figure 7.3 A simple pseudocode for applying genetic algorithm to optimize solar radiation data forecasting.

to determine which individuals to add to the population. Over time, the population evolved to contain better candidate solutions. After a specified number of generations, the algorithm returned the best candidate solution. The importance of using meta-heuristic algorithms to optimize PV power forecasting models [50] cannot be overstated. Accurate forecasting is crucial for effective energy management, as it can lead to significant cost savings and reduced carbon emissions. The use of these algorithms can help achieve higher forecasting accuracy, reduce errors, and improve the overall performance. Thus, they are crucial for the development of sustainable energy systems.

Overall, the literature review suggests that AI techniques have shown great potential for modeling and estimating solar radiation. AI algorithms, such as ANNs, SVMs, fuzzy logic, and DL algorithms, provide accurate and efficient models for estimating solar radiation. However, there is still scope for advancement, and more studies are required to create AI frameworks for solar radiation predictions that are more reliable and precise. A straightforward pseudocode for using an SVM model to improve solar energy predictions is shown in Figure 7.4. To assess the success of the model, the data were initially separated into training and testing groups. The data were then normalized to have a mean of zero and a variation of one, enhancing the performance of the SVM model. The hyperparameters and kernel function used to determine the SVM model can affect how well the model fits data. After training on the training set, the SVM model was tested on the testing set using a defined measure, such as the mean absolute error or root mean square error.

To find the best model, different ways of combining hyperparameters can be attempted. The top-performing SVM model was then returned. To

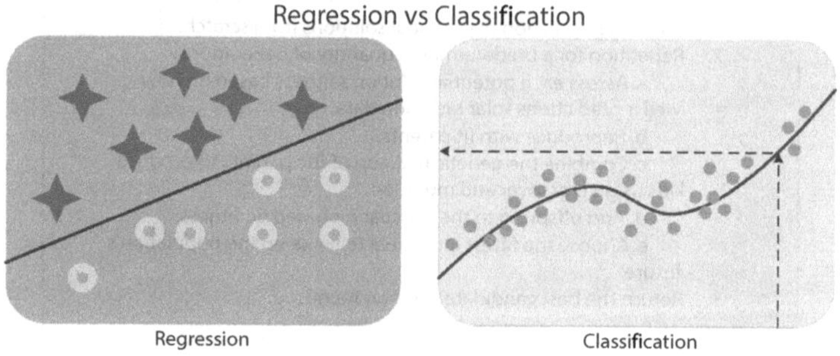

Figure 7.4 Classification and regression through SVM.

1. Create training and test groups from the data.
2. Normalize the data to have zero mean and unit variance.
3. Define the SVM model with a specified kernel function and hyperparameters.
4. Train the SVM model on the training set.
5. Utilize the stated metric to assess the model's success on the testing collection.
6. Repeat steps 3-5 with different combinations of hyperparameters to find the best performing model.
7. Return the best performing SVM model.

Figure 7.5 A simple pseudocode for applying SVM model to optimize solar radiation data forecasting.

assess the efficacy of the model, the data were first divided into training/test groups (see Figure 7.4), which is a pseudo-code. The data are then normalized to have a mean of zero and a variation of one, which can enhance the performance of the SVM model. The hyperparameters and kernel function used to determine the SVM model can have an impact on how well the model fits the data. In the following step, the SVM model is taught on the training set, and its performance is assessed on the testing set using a predetermined measure such as the average absolute error/RMSE. Various hyperparameter combos can be tested to identify the best-performing model. As shown in Figure 7.5, support vector machines (SVMs) can be applied for categorization, regression, and outlier identification. SVMs can divide data into different groups by establishing a decision limit that maximizes the difference between two classes, which makes them especially useful when working with high-dimensional data. Given a training dataset consisting of feature vectors $x_1, x_2, \ldots, x_n$ and the corresponding class labels $y_1, y_2, \ldots, y_n$ (where y_i is either +1 or −1), we want to find a hyperplane in the feature space that separates the two classes with the largest margin. The hyperplane in the SVMs is mathematically defined in Eq. (7.1).

$$w \wedge Tx + b = 0 \tag{7.1}$$

where the weight vector is denoted by w, bias by b, and transpose by T. Margin is the distance of the hyperplane from the nearest data points of each class. If x_+ and x_- are the closest positive and negative data points, respectively, then the margin is $2/|(|w|)|$, where $|(|w|)|$ is $L2$ norm of the weight vector. To find the hyperplane that maximizes the margin, we can solve the following optimization problem: maximize $\frac{1}{2}|(|w|)|\wedge 2$, subject to Eq. (7.2)

$$y_i(w^T x_i + b) \geq 1 \, for \; i = 1,2,\ldots,n \tag{7.2}$$

Once the hyperplane has been found, new data points can be classified by evaluating the sign of the function using Eq. (7.3)

$$f(x) = w \wedge T\; x + b \tag{7.3}$$

The point is categorized as belonging to either the positive or negative class, depending on whether *f(x)* is positive or negative.

B. *Deep learning (DL)* involves numerous neural network layers that are constructed and trained to perform certain tasks. Convolutional neural networks (CNNs) and recurrent neural networks (RNNs) are two of the most frequently employed deep neural networks.

a) *CNNs* are a special type of neural network that is excellent for tasks involving the classification of images and videos. Convolutional layers identify spatial trends in the input data.

Given an input picture I, feature maps are first constructed using a sequence of convolutional filters. A small matrix of weights, known as a convolutional filter, was applied to a small area of the input image. The output of the convolutional layer is governed by Eq. (7.4).

$$h(x,y,f) = relu\left(\sum_i \sum_j \sum_c w_i, j, c, f * I(x+i, y+j, c) + b_f\right) \tag{7.4}$$

where $h(x,y,f)$ is the output of the f-th filter at position $(x,y), w_i, j, c, f$ is the weight at position (i,j) in the filter for channel c and filter

$f, I(x+i, y+j, c)$ is the input image value at position $(x+i, y+j)$ in channel c, b_f is the bias term for the *f*th filter, and relu() is the rectified linear activation function. The result from the convolutional layer is then allowed through one or more fully connected layers, which performs a nonlinear transformation of the feature maps to generate the final output.

b) For sequential data such as time series or natural language text, RNNs are an effective neural network type. They used recurrent layers to learn the temporal patterns in the input data.

Given a sequence of inputs $x_1, x_2, \ldots, x_T$, we want to predict a sequence of outputs $y_1, y_2, \ldots, y_T$. For the RNN, the hidden state at time t is h_t as shown in Eq. (7.5). h_t is functionally dependent on $h_{\{t-1\}}$ and x_t, where $h_{\{t-1\}}$ is the previous hidden state, and x_t is the current input:

$$h_t = f(W_{xh}x_t + W_{hh}h_{\{t-1\}} + b_h) \tag{7.5}$$

where W_{xh} and W_{hh} are weight matrices, b_h is a bias vector, and $f()$ is an activation function, such as a hyperbolic tangent or sigmoid function. The output y_t at each time step was then computed as a function of the current hidden state:

$$y_t = g(W_{hy}h_t + b_y) \tag{7.6}$$

where W_{hy} is a weight matrix, b_y is a bias vector, and $g()$ is an activation function such as the softmax function for classification tasks or the identity function for regression tasks.

C. *ANN*, a computational model based on brain anatomy and function [51], consists of interconnected processing units called neurons that are organized into layers. ANNs are trained by using large amounts of data to identify patterns and relationships. Subsequently, they are applicable to many different applications, including categorization, regression, and forecasting. The neurons in an ANN process information by receiving inputs, performing calculations, and producing output. The connections between the neurons are assigned weights, which are modified over the course of training to improve the performance of the network. The basic equation for an ANN is the activation function $f_{act}()$, which determines the output of a neuron based on its input. Here, the inputs are combined with the weights of the connections to the neuron, and then transformed by $f_{act}()$, to produce the output.

A simple example of $f_{act}()$ is the binary threshold function, which produces an output of either 0 or 1 depending on whether the input is below or above a threshold value. A commonly used $f_{act}()$ in ANNs is the sigmoid function, which maps any input to the range $(0,1)$. The activation function is mathematically represented in Eq. (7.7).

$$y = f_{act}(x * w + b) \tag{7.7}$$

where y is the output of the neuron and $f_{act}()$ is an activation function. The input vector to the neuron is denoted by x, the weight vector of the connections to the neuron by w, and b is the bias, which is a scalar value added to the weighted sum of the inputs. The specific form of $f_{act}()$ and the weights as well as the biases are established during the training phase, when the ANN is given a series of feeds and expected results, and the biases and the weights are changed to reduce the error between the desired and actual outputs.

D. Fuzzy logic is an analytical framework for handling ambiguity and imprecision. Classical logic, also called "crisp logic," is based on true/false values. Fuzzy logic, however, is based on membership values between 0 and 1. Fuzzy logic can be used to represent and process uncertain or vague information, such as human knowledge or natural language expressions, in a more flexible and intuitive manner. In fuzzy logic, a concept is described by a fuzzy set, which assigns a membership degree to each possible input value. Fuzzy inference involves the application of a set of rules to the input values to deduce a conclusion represented by a fuzzy output. The output can then be defuzzied or transformed into a single crisp value to produce the final decision or prediction. Fuzzy logic has applications in a wide range of fields [29–31], including control systems, image and speech processing, decision-making, and artificial intelligence.

E. The principles of natural selection and genetics serve as inspiration for a search-based optimization algorithm known as the *genetic algorithm (GA)*. This is used to solve problems that can be represented as a search space for possible solutions. The algorithm encodes candidate solutions as a set of parameters or variables, called a chromosome, and represents a population of these solutions. Using selection, crossover, and mutation, the genetic algorithm simulates natural selection to produce new individuals from an existing population. The fitness of each solution was evaluated and the fittest solutions were selected to produce the next generation. This procedure is iterated across several generations until a workable answer is discovered or a stopping requirement is satisfied. Problems that are challenging or impracticable to address using conventional optimization techniques can benefit from the use of gas [5]. These issues may involve a large number of variables, difficult or nonlinear goals, or erratic or noisy data. They have applications in areas such as engineering, finance, and bioinformatics. The mathematical representation of a genetic algorithm can be summarized as in Eq. (7.8) where $X(t)$ is the population of chromosomes in generation t, $SELECT$ is the selection operator, $EVAL$ is the evaluation function, $CROSSOVER$ is the crossover operator, $MUTATE$ is the mutation operator, $f(x)$ is the fitness function, p_c is the crossover probability, and p_m is the mutation probability.

$$X(t+1) = SELECT\left(EVAL\left(X(t)\right), f(x)\right) + CROSSOVER(X(t), p_c) + MUTATE(X(t), p_m) \tag{7.8}$$

F. An expert system (ES) simulates a human expert's decision-making process in a certain field. It is intended to offer clear answers and advice to users by combining knowledge from various sources with reasoning and inference techniques. Expert systems use a combination of artificial intelligence, machine learning, and expert knowledge to solve problems and to make decisions in a particular field. They are often used in domains where there is a high degree of uncertainty or where human expertise is limited, and are designed to augment, rather than replace, human decision-making.

An expert system in photovoltaic (PV) forecasting [52–55] uses AI and ML algorithms to predict the future performance of a photovoltaic power system. To calculate the anticipated power generated by a PV system, it uses previous data, as well as weather forecasts, and makes recommendations on how to optimize its performance. The goal of an expert system in PV forecasting [52] is to increase the predictability of energy generation and help optimize the performance of a PV system in real time.

G. Hybrid Intelligent Systems

A hybrid intelligent system combines different AI techniques to solve complex problems. It integrates different intelligent methodologies, such as expert systems [52–55], fuzzy logic [29–31], neural networks, and evolutionary algorithms, to provide a more comprehensive solution than any technique alone. The goal of a hybrid intelligent system is to exploit the strengths of each technique to overcome their individual limitations and provide a more robust solution. Hybrid intelligent systems are often used in applications where the problem is too complex for a single technique to handle, and where multiple sources of information and decision-making approaches are needed to arrive at a satisfactory solution.

7.3 Modeling and Prediction of Data on Solar Irradiance Using AI Approaches

Several artificial intelligence (AI) approaches [56] have been used to model and predict solar radiation data, including:

A. ANNs are popular AI techniques for solar radiation forecasting [57–59]. They are trained [29] using historical solar radiation data and then used to predict the solar radiation for a given location and time. The intricate interactions involving solar radiation and environmental factors, including temperature, humidity, atmospheric pressure, and cloud cover can be modeled using ANNs. Models such as multilayer perceptron (MLP), radial basis function (RBF), and general regression neural network (GRNN) have been used to forecast solar radiation [57–59].

B. In general, SVM models are used to model complex relationships [62–64] between solar radiation and various environmental factors [60, 61], such as temperature [62], humidity [60], and atmospheric pressure. In the SVM models, a decision boundary is set up to divide the solar radiation data into positive and negative samples in a high-dimensional feature space. This boundary was then used to predict the solar radiation for a given location and time. SVMs are particularly well suited for regression and classification problems, where the data have a high number of features or are not linearly separable.

These models can be trained using historical solar radiation data [61], weather patterns, and other relevant factors to predict future solar radiation levels. The algorithm generates a hyperplane that separates the data into classes, where each class corresponds to a different solar-radiation level. In Figure 7.6, the data points nearest to the hyperplane serve as support vectors, which are vector representations of the input space.

The hyperplane is the boundary that separates the input vectors into two classes: one representing high solar radiation and the other representing low solar radiation. The location and orientation of the hyperplane in

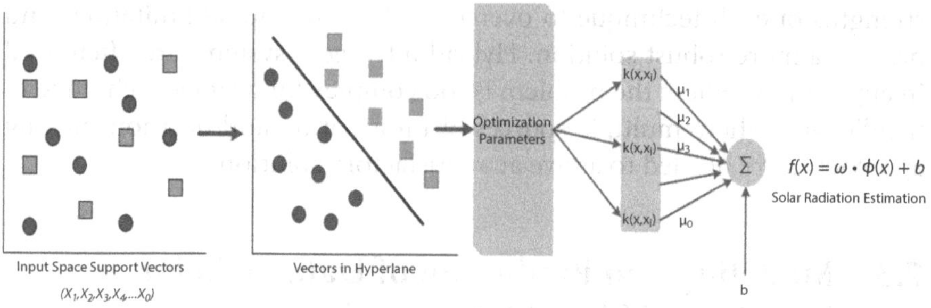

Figure 7.6 Solar radiation estimation using SVM.

the input space were determined using support vectors. The SVM method determines the hyperplane with the largest difference between support vectors for each class. By maximizing the margin, the SVM algorithm ensures that the hyperplane generalizes well to the new data points. By using the input space and support vectors to find the hyperplane, the SVM ensures accurate predictions and efficient renewable energy production. Prediction is based on the class to which a new instance of data belongs.

C. Linear regression is a simple but powerful method for predicting the solar irradiance. This involves fitting a straight line to the data points and using the slope and intercept of the line to make predictions. The formula for a linear regression model is $y = mx + b$, where y is the predicted value, x is the input variable (e.g., time of day), m is the slope, and b is the intercept.

D. Decision trees are a supervised learning algorithm that can be used to predict solar irradiance. They divided the data recursively into subgroups according to the value of a selected characteristic until each subset included only one class. The formula for the decision tree model is given in Eq. (7.9).

$$f(x) = \Sigma c_i * I(x \in R_i) \tag{7.9}$$

where f(x) is the forecasted value, c_i is the average value of the target variable in the *i*th region R_i, and $I(x \in R_i)$ is an indicator function that returns 1 if x belongs to R_i and 0 otherwise.

E. Multiple decision trees are combined in *random forests*, an ensemble learning technique, to increase prediction accuracy. On a large number of randomly chosen data subsets, they built decision trees and then aggregated their predictions. The formula for the random forest model is shown in Eq. (7.10).

$$f(x) = \frac{1}{n}\sum_{i=1}^{n} f_i(x) \tag{7.10}$$

where $f_i(x)$ is the prediction of the *i*th decision tree and the count of trees is denoted by n.

F. Fuzzy logic incorporates uncertainty in the input variables, making it well-suited for modeling real-world problems, such as solar radiation prediction. Fuzzy models use linguistic variables, such as "high," "medium," and "low," to represent the input variables, and fuzzy rules are established to map these linguistic variables to the output variable, solar radiation. The concept behind fuzzy logic is to represent the relationships between different inputs and outputs as fuzzy sets, which are sets with degrees of membership that can be anywhere between 0 and 1. In the context of solar radiation prediction, fuzzy logic can be used to model the relationship [65] between different weather variables [66] and solar radiation levels. The input variables can be described by fuzzy sets [29] and the output can be computed using a set of fuzzy rules that capture the relationships between the input and output variables. Fuzzy logic has been applied to various solar radiation prediction problems [29–31, 67, 68], and some studies have shown that it can provide accurate results [68].

G. Deep learning (DL): Deep learning algorithms [32–38], including convolutional neural networks (CNNs) and recurrent neural networks (RNNs), have been used to model spatial and temporal patterns in solar radiation data. CNNs are well suited for modeling spatial patterns because they can capture local patterns in the data. In contrast, RNNs can capture temporal patterns by modeling data sequences.

H. Hybrid models: Combining different AI approaches such as ANNs and fuzzy logic can lead to more accurate and robust solar radiation prediction models [49]. Hybrid models [27, 41] can take advantage of the strengths of different AI techniques and provide more accurate predictions than standalone models.

7.4 Detailed Comparative Analysis of Different AI Approaches Used in Modeling and Forecasting of Data on Solar Radiation

The choice of an appropriate AI method for the simulation and prediction of data on solar radiation depends on the specific requirements of the application. In this section, we provide a comparative analysis of the different AI techniques used in solar radiation forecasting, including ANNs, SVMs, fuzzy logic, decision trees, deep learning (DL), and hybrid models.

ANNs can be trained using historical solar radiation data and then used to predict the solar radiation for a given location and time. Different types of ANN models have been used for solar radiation prediction, and the choice of the appropriate model depends on the complexity of the problem and available data. Because AI solutions have been proven to enhance the effectiveness and reliability of solar PV systems, their application in the solar sector has been continuously increasing. Figure 7.7 shows the data on the market share of different AI technologies in the solar PV system as well as the corresponding year range, indicating a dynamic landscape of AI adoption in the industry. During 2003–2007, ANNs were the dominant technology, accounting for 25% of the market share, followed by support vector machines (SVMs) with 20%, and fuzzy logic with 10%. The 2008–2012 period saw the emergence of genetic algorithms (GAs), accounting for 5% of market share. From 2013 to 2017, deep learning gained the

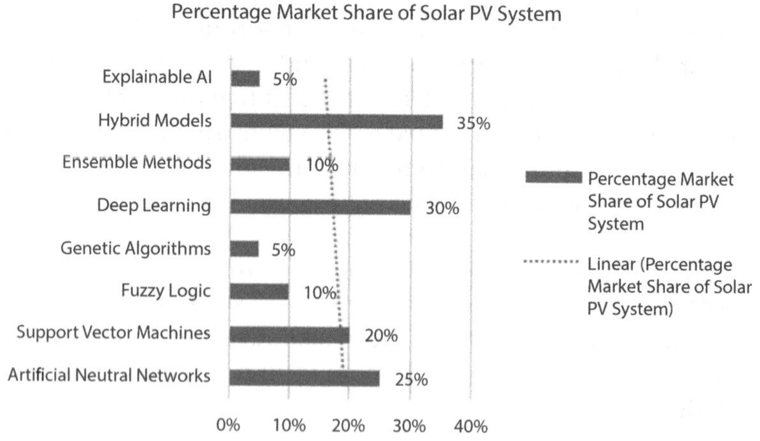

Figure 7.7 Overall percentage in market share of solar PV system for different AI technologies at different time periods.

highest market share (30%), followed by the ensemble methods (10%). The period 2018–2022 witnessed the dominance of hybrid models, accounting for 35% of the market share, with explainable AI accounting for only 5%.

These percentages indicate a shift in preference from traditional AI technologies such as ANNs, SVMs, and fuzzy logic to newer and more sophisticated algorithms, such as deep learning and hybrid models. It is worth noting that the market share of AI technologies in the solar industry is continuously evolving and can be influenced by various factors, such as advancements in technology, changes in regulations, and market demands. As such, the industry can expect to see further advancements and changes in the market share of AI technologies in the coming years.

SVM models are used to model the complex relationships between solar radiation and various environmental factors, such as temperature, humidity, and atmospheric pressure. In SVM models, the solar radiation data are mapped into a high-dimensional feature space, where a decision boundary is established to separate the positive and negative samples. This boundary is then used to predict solar radiation for a given location and time. SVM models have been found to be effective in solar radiation prediction; however, they require a high-dimensional feature space and may not perform well when the data are highly nonlinear. ANNs and SVMs seem to perform the best, according to a study [28], in part because of their ability to rapidly adapt to various external conditions. Fuzzy models [29–31, 67, 68] incorporate uncertainty in the input variables, making them well suited for modeling real-world problems, such as solar radiation prediction.

Fuzzy models use linguistic variables, such as "high," "medium," and "low," to represent the input variables, and fuzzy rules are established to map these linguistic variables to the output variable, solar radiation. Fuzzy models have been found to be effective in solar radiation prediction; however, they require expert knowledge to establish fuzzy rules.

In the context of solar radiation prediction, fuzzy logic can be used to model the relationship [65] between weather variables [66] such as temperature, humidity, cloud cover, and solar radiation levels. The input variables can be described by fuzzy sets [29] and the output can be computed using a set of fuzzy rules that capture the relationships between the feed-in and feed-out variables. Fuzzy logic has been applied to various solar radiation prediction problems [67, 68], and some studies have shown that it can provide accurate results [68].

CNNs and RNNs are deep-learning algorithms that have been employed to model spatial and temporal patterns in solar radiation data, respectively. CNNs are well-suited for modeling spatial patterns, as they can capture local patterns in the data. RNNs, on the other hand, can capture temporal

Table 7.1 Introduction and acceptance of AI technologies over the years in solar PV forecasting domain.

Technologies	Duration (in years)	Pros	Cons	Level of acceptability
Neural Networks	2003–2007	Laid foundation for future developments	Limited accuracy of solar forecasting	Low
Fuzzy Logic	2003–2007	Focused exploration of traditional methods	Limited availability and sophistication of AI tech.	Low
Support Vector Machines	2008–2012	Increased accuracy of forecasting	Limited access to data	Medium
Decision Trees	2013–2017	Further improvements in accuracy	Difficulty interpreting outputs	High
Random Forests	2013–2017	Greater access to data and computation res.	Potential for overfitting	High
Gradient Boosting	2013–2017	Development of machine learning models	–	Low
Conv. Neural Net.	2018–2022	Promise of accurately predicting solar energy	Require significant computation resources	High
Recurrent Neural Net.	2018–2022	More comprehensive and robust AI models	Lack of interpretability in some models	High
Deep Learning	2018–2022	Focus on deep learning models		High

Table 7.2 A comparative analysis of different AI approaches used in modeling and prediction of data on solar radiation.

Approach	Strengths	Weaknesses	References
Artificial Neural Networks (ANNs)	• Model complex relationships between solar radiation and environmental factors • High accuracy in predictions • Well-established approach with a large body of literature • Can handle large amounts of data and noisy or incomplete data. • Capable of online learning, and can generalize well to new data	• Require large amounts of data for training. • Can be prone to overfitting • Computationally expensive to train. • Difficult to interpret	[25–27, 29, 57–59]
Support Vector Machines (SVMs)	• Good performance in high-dimensional feature spaces. • Robust to noisy data • Solid theoretical foundation • Less prone to overfitting than ANNs, and computationally efficient	• Computationally expensive • The decision boundary may not be representative of the underlying relationship between solar radiation and environmental factors • Sensitive to the choice of kernel function. • Hyperparameters, may not perform well with noisy or incomplete data, and difficult to interpret	[60–64]
Fuzzy Logic	• Incorporates uncertainty in the input variables. • Can handle missing values in the data • Good at handling imprecise or uncertain data. • Can model complex systems with multiple inputs and outputs, and is interpretable	• Limited representation of the relationship between solar radiation and environmental factors • Can be more difficult to interpret compared to other AI approaches • Computationally expensive. • Subjective specification of fuzzy sets and membership functions. • May not be suitable for modeling nonlinear systems	[27, 29, 49, 65–68]

(*Continued*)

Table 7.2 A comparative analysis of different AI approaches used in modeling and prediction of data on solar radiation. (*Continued*)

Approach	Strengths	Weaknesses	References
Deep Learning (DL)	• Can model spatial and temporal patterns in solar radiation data. • High accuracy in predictions • Can handle large amounts of data and noisy or incomplete data. • Can learn complex representations of data. • Can generalize well to new data	• Requires large amounts of data for training • Can be computationally expensive. • Prone to overfitting. • Difficult to interpret	[32–38]
Hybrid Models	• Combines the strengths of different AI approaches. • More accurate and robust predictions compared to standalone models • Can handle complex systems with multiple inputs and outputs, and can be designed to be interpretable	• Can be more complex to develop and interpret compared to standalone models • Can be computationally expensive. • Requires expert knowledge to design and implement effectively. • May not always result in better performance than using a single modeling technique.	[27, 41, 49]

patterns by modeling sequences of data. DL models have been found to be effective for solar radiation prediction, but they require large amounts of data for training and are computationally expensive. Combining different AI techniques, such as ANNs and fuzzy logic, can lead to more accurate and robust solar-radiation prediction models. Hybrid models [27, 29–31, 41, 49, 52–55] can take advantage of the strengths of different AI techniques and provide more accurate predictions than standalone models.

In conclusion, different AI approaches have their advantages and disadvantages, and the choice of approach is problem-specific with available data. A comprehensive study by Wang [8] suggests that a combination of different AI approaches, such as hybrid models, can provide the most accurate and robust predictions for solar radiation data. Table 7.1 gives a comparative analysis of different AI approaches used in the modeling and prediction of data on solar radiation. It should be noted that the choice of the AI approach depends on the specific problem at hand and available data.

Table 7.2 gives a comparative analysis of different AI approaches used in modeling and prediction of data on solar radiation, summarizing the advantages and disadvantages of fuzzy logic, ANNs, deep learning, SVMs, and hybrid models for the modeling and forecasting of solar PV systems.

7.5 Discussion

AI has significantly improved the modeling and prediction of solar radiation by providing more accurate and reliable methods for forecasting solar irradiance. The ways in which AI has improved solar radiation modeling and prediction can be summarized as

a) **Improved accuracy:** AI algorithms such as neural networks, decision trees [48], and support vector machines can learn complex patterns and relationships within the data, leading to more accurate predictions of solar irradiance.
b) **Faster processing:** AI algorithms are designed to handle large volumes of data quickly, allowing more efficient processing and faster prediction times.
c) **Increased precision:** AI algorithms can model and predict solar radiation at high spatial and temporal resolutions, resulting in more precise predictions that can be used for a variety of applications such as solar energy production and weather forecasting.
d) **Reduced errors:** AI algorithms are less prone to human errors and biases that can occur in traditional modeling [5, 69] and prediction methods.

Integration with other data sources: AI algorithms can be integrated with other data sources, such as weather data, satellite imagery, and ground-based measurements, to provide a more comprehensive understanding of solar radiation patterns and trends.

Overall, AI has revolutionized the field of solar radiation modeling and prediction, providing more accurate, reliable, and efficient methods for forecasting solar irradiance. These advances have helped optimize solar energy production, reduce energy costs, and improve the overall sustainability of the energy industry. These AI approaches have been shown to provide more accurate predictions than traditional statistical methods for modeling and predicting solar radiation data. However, further research is required to develop more robust and accurate AI models for solar radiation prediction. In conclusion, AI techniques have been shown to be a promising approach for solar radiation prediction, and their use is likely to continue to grow as researchers develop new methods and improve the existing ones.

7.6 Conclusion

Over time, the integration of artificial intelligence (AI) technologies has significantly improved the accuracy and efficiency of solar forecasting and modeling. One key application of AI in solar energy is through machine learning algorithms that are trained to predict energy generation based on data from a variety of sources, such as weather forecasts, historical energy production, and real-time monitoring of PV systems. These algorithms can learn from past data and make more accurate predictions as more data are collected. In addition to improving forecasting accuracy, AI technologies have enabled the more efficient operation of PV systems. For example, AI algorithms can be used to optimize the positioning of solar panels to minimize shading and maximize energy production. They can also be used to detect and diagnose potential issues with the system, allowing for quick repairs and minimizing downtime. Overall, the integration of AI technologies has significantly improved the efficiency and effectiveness of solar energy systems. With the ability to accurately forecast energy production and optimize system performance, solar energy is becoming a more reliable and viable source of renewable energy for homes and businesses.

References

1. R. York, "Do alternative energy sources displace fossil fuels?," *Nat. Clim. Chang., 2012 26*, vol. 2, no. 6, pp. 441–443, Mar. 2012, doi: 10.1038/nclimate1451.
2. O. Fricko *et al.*, "Prospects for Alternative Energy Sources in Global Energy Sector," *IOP Conf. Ser. Earth Environ. Sci.*, vol. 434, no. 1, p. 012014, Jan. 2020, doi: 10.1088/1755-1315/434/1/012014.
3. M. Balat and H. Balat, "Biogas as a renewable energy sourcea review," *Energy Sources, Part A: Recovery, Utilization and Environmental Effects*, vol. 31, no. 14. Taylor & Francis Group, pp. 1280–1293, Jan. 2009, doi: 10.1080/15567030802089565.
4. Y. Pandey *et al.*, "An Environment Friendly Energy-Saving Dispatch Using Mixed Integer Linear Programming Relaxation in the Smart Grid with Renewable Energy Sources," *Distrib. Gener. Altern. Energy J.*, pp. 1239–1258–1239–1258, May 2022, doi: 10.13052/DGAEJ2156-3306.37414.

5. P. Denholm and R. M. Margolis, "Evaluating the limits of solar photovoltaics (PV) in traditional electric power systems," *Energy Policy*, vol. 35, no. 5, pp. 2852–2861, May 2007, doi: 10.1016/J.ENPOL.2006.10.014.
6. M. A. Husain, A. Jain, A. Tariq, and A. Iqbal, "Fast and precise global maximum power point tracking techniques for photovoltaic system," *IET Renew. Power Gener.*, vol. 13, no. 14, pp. 2569–2579, Oct. 2019, doi: 10.1049/IET-RPG.2019.0244.
7. M. A. Husain, A. Tariq, S. Hameed, M. S. Bin Arif, and A. Jain, "Comparative assessment of maximum power point tracking procedures for photovoltaic systems," *Green Energy Environ.*, vol. 2, no. 1, pp. 5–17, Jan. 2017, doi: 10.1016/J.GEE.2016.11.001.
8. H. Wang *et al.*, "Taxonomy research of artificial intelligence for deterministic solar power forecasting," *Energy Convers. Manag.*, vol. 214, no. January, p. 112909, 2020, doi: 10.1016/j.enconman.2020.112909.
9. A. Mellit, S. A. Kalogirou, L. Hontoria, and S. Shaari, "Artificial intelligence techniques for sizing photovoltaic systems: A review," *Renew. Sustain. Energy Rev.*, vol. 13, no. 2, pp. 406–419, Feb. 2009, doi: 10.1016/J.RSER.2008.01.006.
10. M. R. Maghami and A. G. O. Mutambara, "Challenges associated with Hybrid Energy Systems: An artificial intelligence solution," *Energy Rep.*, vol. 9. Elsevier, pp. 924–940, Dec. 01, 2023, doi: 10.1016/j.egyr.2022.11.195.
11. V. S. B. Kurukuru, A. Haque, M. A. Khan, S. Sahoo, A. Malik, and F. Blaabjerg, "A Review on Artificial Intelligence Applications for Grid-Connected Solar Photovoltaic Systems," *Energies, 2021, Vol. 14, Page 4690*, vol. 14, no. 15, p. 4690, Aug. 2021, doi: 10.3390/EN14154690.
12. M. A. Husain, A. Jain, and A. Tariq, "A novel fast mutable duty (FMD) MPPT technique for solar PV system with reduced searching area," *J. Renew. Sustain. Energy*, vol. 8, no. 5, p. 054703, Sep. 2016, doi: 10.1063/1.4963314.
13. S. Panda, R. K. Dhaka, B. Panda, A. Pradhan, C. Jena, and L. Nanda, "A review on application of Machine Learning in Solar Energy Photovoltaic Generation Prediction," *Proc. Int. Conf. Electron. Renew. Syst. ICEARS 2022*, pp. 1180–1184, 2022, doi: 10.1109/ICEARS53579.2022.9752404.
14. L. B. Bosman, W. D. Leon-Salas, W. Hutzel, and E. A. Soto, "PV System PredictiveMaintenance: Challenges, Current Approaches, and Opportunities," *Energies, 2020, Vol. 13, Page 1398*, vol. 13, no. 6, p. 1398, Mar. 2020, doi: 10.3390/EN13061398.
15. A. Gregg, T. Parker, and R. Swenson, "A 'real world' examination of PV system design and performance," *Conf. Rec. IEEE Photovolt. Spec. Conf.*, pp. 1587–1592, 2005, doi: 10.1109/PVSC.2005.1488448.
16. S. Ozdemir and G. Sahin, "Multi-criteria decision-making in the location selection for a solar PV power plant using AHP," *Measurement*, vol. 129, pp. 218–226, Dec. 2018, doi: 10.1016/J.MEASUREMENT.2018.07.020.
17. A. Tuohy *et al.*, "Solar Forecasting: Methods, Challenges, and Performance," *IEEE Power Energy Mag.*, vol. 13, no. 6, pp. 50–59, Nov. 2015, doi: 10.1109/MPE.2015.2461351.

18. A. Mellit and S. Kalogirou, "Artificial intelligence and internet of things to improve efficacy of diagnosis and remote sensing of solar photovoltaic systems: Challenges, recommendations and future directions," *Renew. Sustain. Energy Rev.*, vol. 143, p. 110889, Jun. 2021, doi: 10.1016/J.RSER.2021.110889.
19. A. Abubakar, C. F. M. Almeida, and M. Gemignani, "Review of Artificial Intelligence-Based Failure Detection and Diagnosis Methods for Solar Photovoltaic Systems," *Mach. 2021, Vol. 9, Page 328*, vol. 9, no. 12, p. 328, Dec. 2021, doi: 10.3390/MACHINES9120328.
20. A. Mellit, "Recent Applications of Artificial Intelligence in Fault Diagnosis of Photovoltaic Systems," in *Advanced Structured Materials*, vol. 128, Springer, 2020, pp. 257–271.
21. A. A. Al-Katheri, E. A. Al-Ammar, M. A. Alotaibi, W. Ko, S. Park, and H. J. Choi, "Application of Artificial Intelligence in PV Fault Detection," *Sustainability, 2022, Vol. 14, Page 13815*, vol. 14, no. 21, p. 13815, Oct. 2022, doi: 10.3390/ SU142113815.
22. R. Platon, J. Martel, N. Woodruff, and T. Y. Chau, "Online Fault Detection in PV Systems," *IEEE Trans. Sustain. Energy*, vol. 6, no. 4, pp. 1200–1207, Oct. 2015, doi: 10.1109/TSTE.2015.2421447.
23. A. Mellit and S. A. Kalogirou, "A survey on the application of artificial intelligence techniques for photovoltaic systems," in *McEvoy's Handbook of Photovoltaics: Fundamentals and Applications*, Academic Press, 2018, pp. 735–761.
24. S. Kumar, A. Agarwal, and A. Kumar, "Financial viability assessment of concentrated solar power technologies under Indian climatic conditions," *Sustain. Energy Technol. Assessments*, vol. 43, p. 100928, Feb. 2021, doi: 10.1016/J.SETA.2020.100928.
25. A. R. Pazikadin, D. Rifai, K. Ali, M. Z. Malik, A. N. Abdalla, and M. A. Faraj, "Solar irradiance measurement instrumentation and power solar generation forecasting based on Artificial Neural Networks (ANN): A review of five years research trend," *Sci. Total Environ.*, vol. 715, p. 136848, May 2020, doi: 10.1016/J.SCITOTENV.2020.136848.
26. G. Notton, C. Voyant, A. Fouilloy, J. L. Duchaud, and M. L. Nivet, "Some Applications of ANN to Solar Radiation Estimation and Forecasting for Energy Applications," *Appl. Sci., 2019, Vol. 9, Page 209*, vol. 9, no. 1, p. 209, Jan. 2019, doi: 10.3390/APP9010209.
27. K. S. Garud, S. Jayaraj, and M. Y. Lee, "A review on modeling of solar photovoltaic systems using artificial neural networks, fuzzy logic, genetic algorithm and hybrid models," *Int. J. Energy Res.*, vol. 45, no. 1, pp. 6–35, 2021, doi: 10.1002/er.5608.
28. U. K. Das *et al.*, "Forecasting of photovoltaic power generation and model optimization: A review," *Renew. Sust. Energ. Rev.*, vol. 81. Pergamon, pp. 912–928, Jan. 01, 2018, doi: 10.1016/j.rser.2017.08.017.
29. D. Patel, S. Patel, P. Patel, and M. Shah, "Solar radiation and solar energy estimation using ANN and Fuzzy logic concept: A comprehensive and

systematic study," *Environ. Sci. Pollut. Res.*, vol. 29, no. 22. Springer Science and Business Media Deutschland GmbH, pp. 32428–32442, May 01, 2022, doi: 10.1007/s11356-022-19185-z.

30. O. Kisi, "Modeling solar radiation of Mediterranean region in Turkey by using fuzzy genetic approach," *Energy*, vol. 64, pp. 429–436, Jan. 2014, doi: 10.1016/J.ENERGY.2013.10.009.
31. A. Teke, H. B. Yildirim, and Ö. Çelik, "Evaluation and performance comparison of different models for the estimation of solar radiation," *Renew. Sustain. Energy Rev.*, vol. 50, pp. 1097–1107, Oct. 2015, doi: 10.1016/J.RSER.2015.05.049.
32. G. M. Tina, C. Ventura, S. Ferlito, and S. De Vito, "A state-of-art-review on machine-learning based methods for PV," *Appl. Sci.*, vol. 11, no. 16, 2021, doi: 10.3390/app11167550.
33 A. Gensler, J. Henze, B. Sick, and N. Raabe, "Deep Learning for solar power forecasting - An approach using AutoEncoder and LSTM Neural Networks," *2016 IEEE Int. Conf. Syst. Man, Cybern. SMC 2016 - Conf. Proc.*, pp. 2858–2865, Feb. 2017, doi: 10.1109/SMC.2016.7844673.
34. M. Elsaraiti and A. Merabet, "Solar Power Forecasting Using Deep Learning Techniques," *IEEE Access*, vol. 10, pp. 31692–31698, 2022, doi: 10.1109/ACCESS.2022.3160484.
35. G. Alkhayat and R. Mehmood, "A review and taxonomy of wind and solar energy forecasting methods based on deep learning," *Energy AI*, vol. 4, p. 100060, Jun. 2021, doi: 10.1016/J.EGYAI.2021.100060.
36. H. Zhou, Q. Liu, K. Yan, and Y. Du, "Deep Learning Enhanced Solar Energy Forecasting with AI-Driven IoT," *Wirel. Commun. Mob. Comput.*, vol. 2021, 2021, doi: 10.1155/2021/9249387.
37. S. Psomopoulos *et al.*, "SENERGY: A Novel Deep Learning-Based Auto-Selective Approach and Tool for Solar Energy Forecasting," *Energies, 2022, Vol. 15, Page 6659*, vol. 15, no. 18, p. 6659, Sep. 2022, doi: 10.3390/EN15186659.
38. J. F. Torres, A. Troncoso, I. Koprinska, Z. Wang, and F. Martínez-Álvarez, "Big data solar power forecasting based on deep learning and multiple data sources," *Expert Syst.*, vol. 36, no. 4, p. e12394, Aug. 2019, doi: 10.1111/EXSY.12394.
39. H. Malik, A. Iqbal, P. Joshi, S. Agrawal, and F. I. Bakhsh, Eds., "Metaheuristic and Evolutionary Computation: Algorithms and Applications," vol. 916, 2021, doi: 10.1007/978-981-15-7571-6.
40. M. Naseem *et al.*, "A Spider Monkey Optimization Based Global Maximum Power Point Tracking Technique for Photovoltaic Systems," *2022 2nd Int. Conf. Emerg. Front. Electr. Electron. Technol. ICEFEET 2022*, 2022, doi: 10.1109/ICEFEET51821.2022.9847903.
41. A. Asrari, T. X. Wu, and B. Ramos, "A Hybrid Algorithm for Short-Term Solar Power Prediction - Sunshine State Case Study," *IEEE Trans. Sustain. Energy*, vol. 8, no. 2, pp. 582–591, Apr. 2017, doi: 10.1109/TSTE.2016.2613962.

42. L. L. Jiang, D. L. Maskell, and J. C. Patra, "A novel ant colony optimization-based maximum power point tracking for photovoltaic systems under partially shaded conditions," *Energy Build.*, vol. 58, pp. 227–236, 2013, doi: 10.1016/J.ENBUILD.2012.12.001.
43. L. L. Li, S. Y. Wen, M. L. Tseng, and C. S. Wang, "Renewable energy prediction: A novel short-term prediction model of photovoltaic output power," *J. Clean. Prod.*, vol. 228, pp. 359–375, Aug. 2019, doi: 10.1016/J.JCLEPRO.2019.04.331.
44. M. Talaat, T. Said, M. A. Essa, and A. Y. Hatata, "Integrated MFFNNMVO approach for PV solar power forecasting considering thermal effects and environmental conditions," *Int. J. Electr. Power Energy Syst.*, vol. 135, p. 107570, Feb. 2022, doi: 10.1016/J.IJEPES.2021.107570.
45. D. Pattanaik, S. Mishra, G. P. Khuntia, R. Dash, and S. C. Swain, "An innovative learning approach for solar power forecasting using genetic algorithm and artificial neural network," *Open Eng.*, vol. 10, no. 1, pp. 630–641, Jan. 2020, doi: 10.1515/ENG-2020-0073/MACHINEREADABLECITATION/RIS.
46. M. Ratshilengo, C. Sigauke, and A. Bere, "Short-Term Solar Power Forecasting Using Genetic Algorithms: An Application Using South African Data," *Appl. Sci., 2021, Vol. 11, Page 4214*, vol. 11, no. 9, p. 4214, May 2021, doi: 10.3390/APP11094214.
47. N. Zhang, P. K. Behera, and C. Williams, "Solar radiation prediction based on particle swarm optimization and evolutionary algorithm using recurrent neural networks," *SysCon 2013 - 7th Annu. IEEE Int. Syst. Conf. Proc.*, pp. 280–286, 2013, doi: 10.1109/SYSCON.2013.6549894.
48. S. N. Shorabeh, N. N. Samany, F. Minaei, H. K. Firozjaei, M. Homaee, and A. D. Boloorani, "A decision model based on decision tree and particle swarm optimization algorithms to identify optimal locations for solar power plants construction in Iran," *Renew. Energy*, vol. 187, pp. 56–67, Mar. 2022, doi: 10.1016/J.RENENE.2022.01.011.
49. N. A. Abdullah, N. A. Rahim, C. K. Gan, and N. N. Adzman, "Forecasting Solar Power Using Hybrid Firefly and Particle Swarm Optimization (HFPSO) for Optimizing the Parameters in a Wavelet Transform-Adaptive Neuro Fuzzy Inference System (WT-ANFIS)," *Appl. Sci., 2019, Vol. 9, Page 3214*, vol. 9, no. 16, p. 3214, Aug. 2019, doi: 10.3390/APP9163214.
50. M. Naseem *et al.*, "Assessment of Meta-Heuristic and Classical Methods for GMPPT of PV System," *Trans. Electr. Electron. Mater., 2021 223*, vol. 22, no. 3, pp. 217–234, Mar. 2021, doi: 10.1007/S42341-021-00306-3.
51. "Introduction to neural networks," doi: 10.1016/S0140-6736(95)91746-2.
52. "Hybrid Intelligent Systems - Larry R. Medsker - Google Books." https://books.google.co.in/books?hl=en&lr=&id=EXngBwAAQBAJ&oi=fnd&pg=PR13&dq=Modeling+and+Estimation+of+Solar+Radiation+Data+Using+Expert+system+and+Hybrid+Intelligent+Systems&ots=GgpqF-

WO-pa-&sig=qkqGqyomAMzq42rGQwQ9FP2JBJI&redir_esc=y#v=onepage&q&f=false (accessed Apr. 05, 2023).
53. A. Khan, J. P. Li, A. U. Haq, I. Memon, and S. H. Patel, "Emotional-physic analysis using multi-feature hybrid classification," *J. Intell. Fuzzy Syst.*, vol. 40, no. 1, pp. 1681–1694, 2021.
54.. J. Reddy, S. Debnath, B. Das, and Jagadish, "Energy and exergy analysis of wavy plate solar air collector using a novel hybrid expert system," *J. Brazilian Soc. Mech. Sci. Eng.*, vol. 41, no. 10, pp. 1–14, Oct. 2019, doi: 10.1007/S40430-019-1901-X/TABLES/4.
55. S. Debnath, J. Reddy, Jagadish, and B. Das, "An expert system-based modeling and optimization of corrugated plate solar air collector for North Eastern India," *J. Brazilian Soc. Mech. Sci. Eng.*, vol. 41, no. 7, pp. 1–18, Jul. 2019, doi: 10.1007/S40430-019-1782-Z/TABLES/9.
56. R. Fulton, D. Fulton, and S. Kaplan, "Artificial Intelligence: Framework of driving triggers to past, present and future applications and influencers of industry sector adoption," *ArXiv*, pp. 127–144, Mar. 2022, doi: 10.5121/CSIT.2022.120613.
57. A. Alcañiz, D. Grzebyk, H. Ziar, and O. Isabella, "Trends and gaps in photovoltaic power forecasting with machine learning," *Energy Reports*, vol. 9. Elsevier, pp. 447–471, Dec. 01, 2023, doi: 10.1016/j.egyr.2022.11.208.
58. J. Liu, W. Fang, X. Zhang, and C. Yang, "An Improved Photovoltaic Power Forecasting Model With the Assistance of Aerosol Index Data," *IEEE Trans. Sustain. Energy*, vol. 6, no. 2, pp. 434–442, Apr. 2015, doi: 10.1109/TSTE.2014.2381224.
59. L. Liu, D. Liu, Q. Sun, H. Li, and R. Wennersten, "Forecasting Power Output of Photovoltaic System Using A BP Network Method," in *Energy Procedia*, 2017, vol. 142, pp. 780–786, doi: 10.1016/j.egypro.2017.12.126.
60. J. Fan, X. Wang, F. Zhang, X. Ma, and L. Wu, "Predicting daily diffuse horizontal solar radiation in various climatic regions of China using support vector machine and tree-based soft computing models with local and extrinsic climatic data," *J. Clean. Prod.*, vol. 248, p. 119264, Mar. 2020, doi: 10.1016/J.JCLEPRO.2019.119264.
61. M. Guermoui, R. Abdelaziz, K. Gairaa, L. Djemoui, and S. Benkaciali, "New temperature-based predicting model for global solar radiation using support vector regression," *https://doi.org/10.1080/01430750.2019.1708792*, vol. 43, no. 1, pp. 1397–1407, 2020, doi: 10.1080/01430750.2019.1708792.
62. Y. Liu, Y. Zhou, Y. Chen, D. Wang, Y. Wang, and Y. Zhu, "Comparison of support vector machine and copula-based nonlinear quantile regression for estimating the daily diffuse solar radiation: A case study in China," *Renew. Energy*, vol. 146, pp. 1101–1112, Feb. 2020, doi: 10.1016/J.RENENE.2019.07.053.
63. A. N. Khan, H. Nazarian, N. A. Golilarz, A. Addeh, J. P. Li, and G. A. Khan, "Brain Tumor Classification Using Efficient Deep Features of MRI Scans and Support Vector Machine," in *2020 17th International Computer Conference on*

Wavelet Active Media Technology and Information Processing, ICCWAMTIP 2020, 2020, pp. 314–318, doi: 10.1109/ICCWAMTIP51612.2020.9317509.
64. A. Barnett *et al.*, "Image Classification using non-linear Support Vector Machines on Encrypted Data," *IACR Cryptol. ePrint Arch.*, p. 857, 2017, [Online]. Available: https://eprint.iacr.org/2017/857.pdf.
65. A. Chugh, P. Chaudhary, and M. Rizwan, "Fuzzy logic approach for short term solar energy forecasting," *12th IEEE Int. Conf. Electron. Energy, Environ. Commun. Comput. Control (E3-C3), INDICON 2015*, Mar. 2016, doi: 10.1109/INDICON.2015.7443206.
66. M. Rizwan, M. Jamil, S. Kirmani, and D. P. Kothari, "Fuzzy logic based modeling and estimation of global solar energy using meteorological parameters," *Energy*, vol. 70, pp. 685–691, Jun. 2014, doi: 10.1016/J.ENERGY.2014.04.057.
67. S. X. Chen, H. B. Gooi, and M. Q. Wang, "Solar radiation forecast based on fuzzy logic and neural networks," *Renew. Energy*, vol. 60, pp. 195–201, Dec. 2013, doi: 10.1016/J.RENENE.2013.05.011.
68. M. Sridharan, "Short review on various applications of fuzzy logic-based expert systems in the field of solar energy," *Int. J. Ambient Energy*, vol. 43, no. 1, pp. 5112–5128, 2021, doi: 10.1080/01430750.2021.1927839.
69. S. Mohanty, P. K. Patra, and S. S. Sahoo, "Prediction and application of solar radiation with soft computing over traditional and conventional approach - A comprehensive review," *Renew. Sust. Energ. Rev.*, vol. 56. Pergamon, pp. 778–796, Apr. 01, 2016, doi: 10.1016/j.rser.2015.11.078.

8

Application of ANN–ANFIS Model for Forecasting Solar Power

Gulnar Perveen[1], Priyanka Anand[2]* and Amod Kumar[1]

[1]*Defence Research & Development Organisation, Metcalfe House Annexe, Delhi, India*
[2]*Dept. of Electronics & Communication Engg., Bhagat Phool Singh Mahila Vishwavidyalaya, Sonipat, Haryana, India*

Abstract
Reliable solar energy forecasting is essential for the effective planning and optimization of solar photovoltaic systems. Because the power output of photovoltaic systems alternates throughout the day and is unpredictable, it is crucial to plan for power generation because the energy grid depends on solar power forecasting. This forecast is difficult and time-consuming because solar power generation is weather-dependent and unpredictable. The present study used intelligent modeling techniques to forecast solar power under diverse weather conditions. The methods used in this research involve artificial neural network (ANN) and adaptive neuro-fuzzy-inference-system (ANFIS)-based models to forecast PV power using meteorological parameters. It is simple to forecast solar power when the sky is clear, and difficulty arises when the sky is clouded. To forecast the power for sunny/clear, hazy, rainy, and cloudy weather for a composite climate zone, current research is being conducted. Models were created by simulating them on the MATLAB platform, and the statistical parameters validated the accuracy of the findings. In this, four models are investigated, and it has been found that, after a thorough investigation, hazy models are more accurate than other produced models.

***Keywords*:** ANFIS, ANN, forecasting, meteorological parameters

**Corresponding author*: anand_priyanka10@yahoo.co.in

Mohammed Aslam Husain, Md Waseem Ahmad, Farhad Ilahi Bakhsh, P. Sanjeevikumar and Hasmat Malik (eds.) Photovoltaic Systems Technology, (151–176) © 2024 Scrivener Publishing LLC

8.1 Introduction

The motivation and significance section, literature survey, and research gap followed by the novelty are the four subsections that make up the introductory section. The specifics for each of these subsections are as follows.

8.1.1 Motivation and Significance

Energy is an essential and fundamental factor in solar-based PV technology sizing and optimization. In the quest for energy surges, there is a substantial need for renewable energy sources (RESs). In the present scenario, renewable energy power generation has gained importance because of ecological factors, such as pollution and global warming [1, 2]. India's potential in the solar PV sector is quite high, and with the support of right policymaking, it soon became a potential player in the solar-energy-driven market. As of 31 August 2022, India had 1,16,077.64 MW_p of installed capacity overall, with the solar energy industry contributing about 51% of that total, as illustrated in Figure 8.1. Nonetheless, there is still a great deal of untapped potential for using these RESs in electricity generation [3, 4].

Most studies have used simulation methodologies, processes, and strategies to investigate RESs for predicting the solar power. The amount of solar radiation that reaches the surface of the globe is non-stationary in nature because of Earth's rotation and revolution. As a result, the generated power becomes unstable and fluctuating, which has a negative impact on the system's performance. Therefore, it is necessary to create novel and

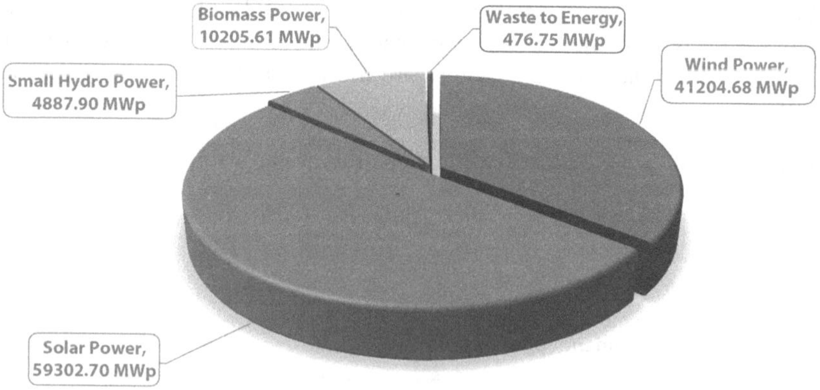

Figure 8.1 Power generation from RESs in India.

creative approaches depending on weather conditions to reduce such oscillations in grid-connected solar PV systems [5, 6].

8.1.2 Literature Survey

According to a thorough literature review, modeling that uses intelligent techniques, such as ANN and ANFIS, is beneficial for generating power, which is trained to overcome the limitations of conventional methodology. Consequently, the ANN–ANFIS model has become useful for resolving complex problems that are otherwise difficult to model and analyze [7, 8].

Demirdelen *et al.* suggested ANN methodology for calculating solar power. In addition, a swarm-based optimization method was used to train the ANN algorithm. The findings suggest that among the numerous approaches employed, the method with the least error is selected [9]. Fara *et al.* proposed an ANN and ARIMA model to forecast the energy generation in solar PV systems. Two case studies are detailed in this chapter: i) Lab BIPV system demonstrated at Bucharest University and ii) a solar PV park located in Romania [10]. Massaoudi *et al.* proposed an effective approach based on an improved Random Forest model. To manage variance optimization for STPF applications and enhance the results, this method makes use of collective attribute selection strategies. The results indicated that the established model is superior to alternative teaching techniques [11]. Mehmet proposed a grey wolf optimizer (GWO), an enhanced type of GWO called I-GWO, for solar PV cells. The double diode model is better than the comparison other models in presenting the physical behaviors of PV cells and modules. It is concluded from the results that I-GWO is more efficient than others for both PV cells and PV modules [12]. For solar photovoltaic systems, Mohammed *et al.* presented a maximum power point tracking (MPPT) method with quick convergence and a smaller search area. One aspect of this includes the tracking's seeking range, and the duty step closes the operating point to the MPP. The study showed that the MPP tracking method is more effective and capable of quick tracking [13].

Naseem *et al.* presented peak-seeking mechanisms for solar PV plants to addressing the energy loss caused by shading. This study compares a variety of traditional and bio-inspired meta-heuristic-based global maximum power point tracking (GMPPT) methods with the fundamentals of partial shade phenomena, PV panel placement, distributed MPPT, and field MPPT, and the results show that the GMPPT approach is the most effective [14]. Mohammed *et al.* suggested methods for tracking the solar array's greatest power point. The examination of various parameters utilizing MPPT approaches is considered and will serve as a useful guide for choosing and comprehending various MPPT methods [15]. Mohammed *et al.* proposed global GMPPT approaches known as fast and intelligent GMPPT (FI-GMPPT) and large and small duty step (LSDS), which sweep the entire P–V curve, while FI-GMPPT further restricts the search area to be covered [16]. Mohammed *et al.* suggested global MPPT method based on the spider monkey optimizing (SMO) method of swarm intelligence-operated optimization. This study concluded that SMO-GMPPT is a quick and reliable GMPPT approach that can follow the real global peak during partial shading phenomena [17]. Faiz *et al.* presented optimization techniques for collecting solar parameters. In this study, a new meta-heuristic optimization technique for collecting diode parameters, that is, the atomic orbital search algorithm, is proposed. This study concludes that the proposed algorithm produces outcomes superior to those of other algorithms [18].

Azka *et al.* proposed a model to predict power of solar system located in tropics. The obtained results revealed that a neural network with two hidden layers achieves more accuracy than a single hidden layer [19]. Ashraf and Chandra proposed an ANN model for predicting the output of solar system which was installed on the island of Lakshadweep. Three models were used in this study, the first of which was a univariate model based on output value and solar radiation. The second model was based on module temp. and solar radiation, whereas the third model comprises the module temp., solar radiation, and clearness index. According to this study, the second model is more accurate [20].

A comprehensive literature analysis reveals that for evaluating complex functions, exactly computing the number of hidden layers and neurons using ANN is a challenging task, as the number is reasonably large. Moreover, the training time required to resolve these complex cases is comparatively high, which slows down the system response. To overcome this drawback, ANFIS models are presented for forecasting solar power.

Leary and Kubby suggested a method for solar power forecasting based on machine learning, classification, and digital image processing. For signal classification and picture identification, input masking and ANN performance tuning strategies were created to increase forecast accuracy. The results demonstrated a better prediction of the target correlation [21]. Akhlaque *et al.* proposed an optimization model of hybrid renewable energy sources (HRESs). Such models are established based on reliability analysis, which includes design factors and cost economics. Several modeling techniques and tools have been established to assist investors in the proper optimal sizing, control, and management approaches of an HRES [22]. Ahmad *et al.* presented modeling for electricity demand whenever there is installation of solar photovoltaic plants. With the inclusion of metaheuristic optimization techniques, the hybrid generation system was modeled to meet the energy demand during peak hours [23]. Akhlaque *et al.* proposed a method for predicting use of all residential structures. However, with power deregulation, load forecasting plays an essential role in the assessment of the measured and forecasted load using an ANN, and the validation of the proposed model is performed using statistical evaluators [24].

For predicting solar PV power, Liu *et al.* suggested an ANFIS model. Several models, including back-propagation, support vector machines (SVM), and radial basis function networks, were proposed in this study. These findings indicate that ANFIS-based models are more accurate than other models [25]. To investigate the impact and influence of climatic elements on forecasting solar energy located in Jordan, Fraihat *et al.* presented the long-short term memory (LSTM) and ANFIS models [26]. Ramadan *et al.* suggested a PV model using ANFIS modeling techniques. This project used the MATLAB platform to train and test several membership functions. The accuracy of the proposed model was confirmed using performance statistical metrics [27]. Gafai *et al.* proposed the ANFIS model, and a comparison was performed with a multi-linear regression (MLR) model to study the performance and accuracy of the proposed model for forecasting solar energy in Nigeria [28]. Diaz *et al.* conducted research based on the ANFIS model and compared its results with multi-linear regression (MLR) and gradient descent optimization (GDO) models using statistical indexes. The results indicate that GDO provides more accuracy than MLR and ANFIS models for Sonora and Mexico City [29].

8.1.3 Research Gap

A thorough analysis of the literature on solar power forecasting reveals that very few studies have elucidated the idea of predicting solar power using intelligent modelling techniques for various weather conditions.

8.1.4 Novelty

The present study novelty lies in developing ANN–ANFIS model to forecast solar power for various weather circumstances, such as sunny/clear, hazy, rainy and foggy with the use of meteorological factors. A composite climate zone-specific model was developed and assessed using statistical evaluation indices.

The organization of the chapter is as follows: *Section 8.2* provides an overview of ANN; the architecture of ANFIS is detailed in *Section 8.3*; *Section 8.4* discusses the characterization of solar plant; classification of weather conditions is detailed in *Section 8.5*; *Section 8.6* involves statistical performance indicators, and the development of ANN–ANFIS models is detailed in *Section 8.7*, followed by the results and discussions in *Section 8.8*.

8.2 Overview of ANN

The multi-layered neural network (ANN) comprises an input layer, hidden layers, and an output layer. When the number of hidden layers increases, the network becomes deeper because each node in one layer is associated with other nodes in the subsequent layer, as depicted in Figure 8.2. A non-linear activation function is applied to the weighted sum of input node inputs before it is passed on to another node in the following layer as its input [30–32].

8.2.1 Models of ANN

There are many types of ANN models:

1. **Multilayer perceptron (MLP):** This feed-forward ANN model, which is appropriate for complicated nonlinear

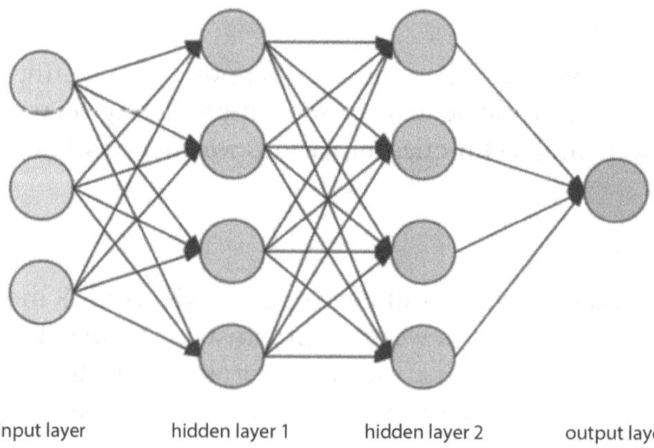

Figure 8.2 Architecture of ANN.

models, maps input data to output data composed of multiple layers, and each layer is connected to the next.
2. **Radial basis function (RBF):** Like MLP, it is a supervised learning network, but it only employs one hidden layer. Hidden neurons in an RBF network coexist near one another but are essentially unconnected.

8.3 ANFIS Architecture

ANFIS is based on the Takagi–Sugeno Fuzzy Inference System, which comprises five layers. The first layer, i.e., the fuzzification layer, receives the input and administers the membership function. Layer 2 is accountable for creating the firing strengths of the rules. Layer 3 normalizes the estimated firing strengths by dividing the value of the overall firing strength. Layer 4 receives the normalized input, and the final output is defuzzified to the last layer, as shown in Figure 8.3 [33, 34].

8.3.1 ANFIS Layers

Specifically, it has the following five layers:

1. **Layer 1:** In this layer, the node performs the role of an adaptive node and provides the membership function degree, as given in Eqs. (8.1) and (8.2), respectively:

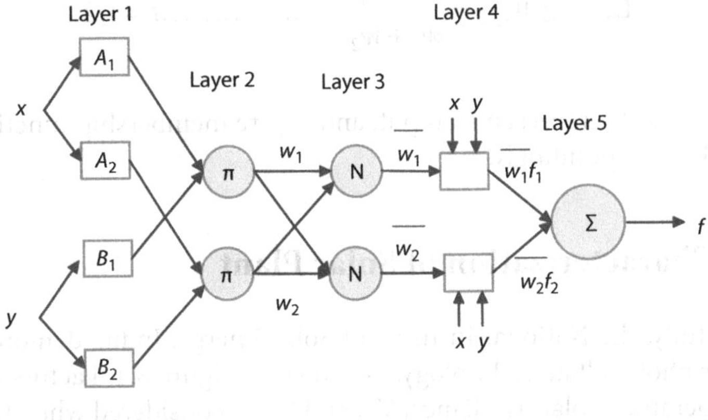

Figure 8.3 ANFIS architecture.

$$L_{1,i} = \mu_{x,i}(v) \ldots\ldots i = 1,2 \tag{8.1}$$

$$L_{1,j} = \mu_{x,j}(v) \ldots\ldots j = 1,2 \tag{8.2}$$

2. **Layer 2:** Here, node is set and is referred to 'Π' with the output being the product of the incoming signals and being represented in Eq. (8.3) as:

$$L_{2,i} = \mu_{x,i}(v) * \mu_{y,j}(v) \ldots\ldots i = 1,2 \tag{8.3}$$

3. **Layer 3:** Here, the node is either set or nonadaptive, which is denoted by the letter 'N' in Eq. (8.4), and denotes normalization:

$$L_{3,i} = \frac{w_1}{w_1 + w_2} \ldots\ldots i = 1,2 \tag{8.4}$$

4. **Layer 4:** Every node contributes an adaptive node to the output to achieve this, as seen in Eq. (8.5):

$$L_{4,i} = \overline{w_i} f_i = \overline{w_i} a_i p + b_i q + c_i) \ldots\ldots\ldots\ldots\ldots i = 1,2 \tag{8.5}$$

5. **Layer 5:** The output layer is composed of a single fixed node with the name 'Σ' that adds the incoming signals, as shown in Eq. (8.6):

$$L_{5,i} = \sum \overline{w_i} f_i = \frac{\Sigma w_1 f_1}{w_1 + w_2} \ldots\ldots\ldots\ldots\ldots i = 1,2 \tag{8.6}$$

where $L_{1,i}$ and $L_{1,j}$ represents output, and $\mu_{x,i}$ are membership function sets A_i and B_i correspondingly.

8.4 Characterization of Solar Plant

In this study, the National Institute of Solar Energy, India, demonstrated the solar photovoltaic technology, as shown in Figure 8.4. Factors such as cell temperature, solar irradiance, V_{oc}, and I_{sc} are considered when training

Figure 8.4 Solar PV technology demonstration at NISE, India.

Table 8.1 Specification of solar plant.

Characteristics	Specs
η_o (Module efficiency)	16.94%
Cell efficiency	18.16%
I_{sc} (Short-circuit current)	8.59A
T_{amb} (Ambient operating temp.)	−2°F to 124°F
V_{oc} (Open-circuit voltage)	51.23 V
N_{OCT} (Normal operating cell temperature)	123.25°F

the proposed model, and the specifications of the 250 W_p polycrystalline solar-based PV technology are listed in Table 8.1.

Further, the data obtained are 5 years of averaged data and collected daily and are presented in Table 8.2 for the composite climate.

8.5 Classification of Weather Condition

According to the incident solar rays and sunlight hours availability, four types of weather conditions are described [35]:

Table 8.2 Measured PV data set for composite climate zone.

Month in an year	Power (W)	V_{OC} (V)	I_{SC} (A)	V_{pmax} (V)	I_{pmax} (A)	Solar irradiance (W/m²)	Cell temp. (°C)
1	59.87	26.04	4.89	22.26	4.65	342.63	25.89
2	65.38	26.89	5.91	21.89	5.25	451.85	34.12
3	62.34	27.23	5.99	21.08	5.99	513.96	36.78
4	59.56	26.89	5.97	21.68	5.50	526.78	37.45
5	56.50	25.54	5.65	19.29	5.37	558.23	39.48
6	51.49	25.44	5.81	19.09	4.96	529.78	47.02
7	44.08	24.57	4.89	19.27	4.36	534.56	41.48
8	46.73	24.68	4.57	19.85	4.67	412.89	42.45
9	48.49	25.84	4.69	19.96	4.96	426.75	40.58
10	63.24	25.86	5.78	19.56	5.85	465.89	43.69
11	50.06	26.38	4.96	20.68	4.76	347.26	44.78
12	57.78	27.83	5.52	20.59	5.12	360.78	29.36
Avg.	**55.46**	**26.09**	**5.38**	**20.43**	**5.12**	**455.94**	**38.59**

1. Sunny

The amount of diffuse solar energy (H_d) was equal to or less than 25% of the amount of global solar energy (H_g), and the number of sunshine hours was at least 9 h.

2. Hazy

If H_d is between 50% and 25% H_g, there is also 7–9 h of sunshine.

3. Rainy

If the sunrays fall within 5–7 h, H_d is lower than 75% or greater than 50% of the H_g.

4. Cloudy

If the duration of sunlight is less than 5 h, H_d is greater than 75% of H_g.

8.6 Statistical Performance Indicators

Performance indicators such as the Mean Absolute Percentage Error (MAPE) and normalized Mean Absolute Error (n-MAE) were considered for computing the model accuracy [36, 37].

8.6.1 MAPE

It is defined as the ratio between the absolute error added together and measured data averaging to fixed percentage as indicated in Eq. (8.7) is written as:

$$\text{MAPE}_{\%} = \frac{1}{n}\sum_{i=1}^{n}\left|\frac{E}{m_i}\right| * 100 \tag{8.7}$$

8.6.2 n-MAE

Given that solar irradiation varies during the day from dawn to dusk and that this varies the power produced by solar PV systems, n-MAE are sought, which are provided by Eq. (8.8) as follows:

$$\text{n-MAE\%} = \frac{1}{n}\sum_{i=1}^{n}\left|\frac{E}{max(m_i)}\right| * 100 \tag{8.8}$$

where n denotes the no. of observations, m_i and f_i represent the observed and forecasted data, and $E = (m_i - f_i)$ denotes the absolute error.

8.7 Development of ANN–ANFIS Model

In this study, ANN–ANFIS models were established for PV power forecasting. These models include the network topology, weighted node connectivity, and input and output selection. A recurrent algorithm for optimizing the hidden layers and nodes in relation to the correctness of the training dataset experimentally drives the entire process of selection and topology. Further, simulating training data and network adaptation, functions namely 'TRAINLM' and 'LEARNGDM' are used which are given by Eq. (8.9)–Eq. (8.10) as [38–41]:

$$P_{PV} = [P_{PV,STC} * \frac{G_T}{1,000} * [1 - \gamma * (T_j - 25)]] * N_{PVs} * N_{PVp} \tag{8.9}$$

$$\text{and } T_j = T_{amb} + \frac{G_T}{800} * (N_{OCT} - 20) \tag{8.10}$$

where $P_{PV,STC}$ stands for PV system power under STC conditions (in Watts), P_{PV} for PV panel output (in Watts), and G_T for solar irradiance at STC (in Watts per square meter); T_{amb} represents amb. temperature (°C); N_{PVS} stands for the PV panel connected in series; γ represents temperature parameter; N_{PVP} stands for PV panel linked in parallel; T_j represents panel temperature (°C), and N_{OCT} was constant.

8.8 Results

ANN–ANFIS models have been established for solar power forecasting in solar systems using input parameters such as cell temperature, solar radiation, and output parameters, namely power, as shown in Figure 8.5. Statistical performance indicators were used to validate the proposed model, as presented in Table 8.3.

Table 8.3: In contrast to the ANN models, which have MAPE and n-MAE values of 0.109% and 0.085%, respectively, the ANFIS models have MAPE values of 0.036% and n-MAE values of 0.038%, which makes them more accurate. Furthermore, the amount of energy produced by solar-based energy systems is highly dependent on changing climatic circumstances; as a result, the current study is expanded to include an analysis of four different weather conditions: sunny, hazy, rainy, and foggy.

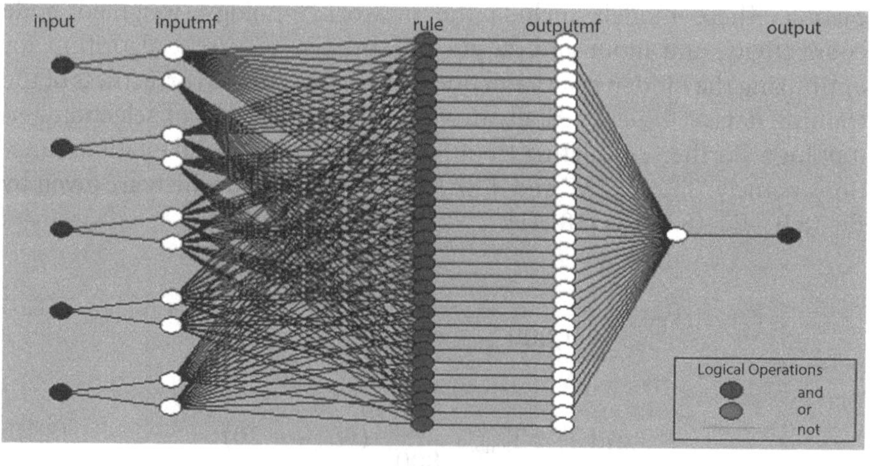

Figure 8.5 Schematic of ANFIS model simulated in MATLAB.

Table 8.3 ANN-ANFIS model for PV power forecasting.

Month in a year	Solar irradiance (W/m²)	Cell temp. (°C)	Power(W)			MAPE (%)		n-MAE (%)	
			Measured	ANN	ANFIS	ANN	ANFIS	ANN	ANFIS
01	342.63	25.89	59.87	58.12	59.86	0.19	0.012	0.08	0.005
02	451.85	34.12	65.38	65.01	65.37	0.12	0.055	0.07	0.012
03	513.96	36.78	62.34	62.15	62.33	0.16	0.094	0.02	0.026
04	526.78	37.45	59.56	58.23	59.50	0.08	0.008	0.06	0.087
05	558.23	39.48	56.50	56.41	56.52	0.09	0.006	0.06	0.021
06	529.78	47.02	51.49	51.02	51.42	0.13	0.031	0.08	0.045
07	534.56	41.48	44.08	43.56	44.01	0.07	0.002	0.10	0.068
08	412.89	42.45	46.73	46.01	46.72	0.08	0.004	0.10	0.043
09	426.75	40.58	48.49	48.02	48.23	0.15	0.069	0.12	0.029
10	465.89	43.69	63.24	63.02	63.26	0.06	0.035	0.06	0.018
11	347.26	44.78	50.06	49.45	50.01	0.17	0.072	0.09	0.037
12	360.78	29.36	57.78	57.23	57.68	0.11	0.045	0.12	0.069
Avg.	**455.94**	**38.59**	**55.46**	**54.85**	**55.41**	**0.109**	**0.036**	**0.085**	**0.038**

8.8.1 Type-a (Sunny) Model

H_d must be less than 25% of H_g, and sunlight obtainability must be greater than 9 h to meet the requirements of this model. These models were created and the outcomes are listed in Table 8.4. Furthermore, Table 8.4 illustrates that the sunny/clear model produces max. power output of 100.3 W_p at 13:00 h. By employing ANN, average MAPE and n-MAE attained are 0.185% and 0.058%; while by using ANFIS, the error has reasonably reduced with MAPE and n-MAE of 0.052% and 0.008%, respectively.

The measured and predicted results for sunny weather are compared in Figure 8.6. The total time duration considered was from morning 5 am to 6 pm. A maximum power of 100.3 W_p was achieved at 13:00 h. Compared to the temperature factor, the time factor is important and is considered to play a significant role in solar radiation. With the Sunny model, the average MAPE using ANN was 0.185%, but it was observed to be 0.052% using ANFIS.

Table 8.4 ANN–ANFIS model for sunny weather.

Time (h)	Solar irradiance (W/m^2)	Cell temp. (°C)	Power (W)			MAPE (%)		n-MAE (%)	
			Measured	ANFIS	ANN	ANFIS	ANN	ANFIS	ANN
5	16.30	20.48	1.03	1.03	1.12	0.078	0.17	0.002	0.051
6	96.47	21.63	4.89	4.89	4.79	0.056	0.13	0.006	0.064
7	255.14	28.74	13.89	13.89	13.74	0.062	0.18	0.004	0.027
8	436.12	36.58	17.45	17.45	17.28	0.047	0.19	0.013	0.034
9	616.59	43.12	28.96	28.96	28.62	0.013	0.16	0.008	0.096
10	749.68	48.78	54.58	54.57	54.45	0.073	0.14	0.037	0.012
11	750.70	48.69	60.35	60.35	60.23	0.030	0.11	0.008	0.079
12	939.39	49.36	80.48	80.48	80.12	0.012	0.15	0.004	0.036
13	931.31	50.23	100.3	0.1 100.2	100.1	0.089	0.23	0.001	0.045
14	821.78	48.69	79.23	79.23	79.36	0.036	0.15	0.003	0.037
15	684.23	46.85	67.45	67.45	67.48	0.010	0.14	0.004	0.071
16	554.96	43.68	42.89	42.85	42.88	0.023	0.14	0.005	0.034
17	343.45	39.35	16.34	16.35	16.35	0.028	0.19	0.002	0.018
18	152.15	35.87	4.09	4.08	4.10	0.069	0.15	0.001	0.096
Avg.	**612.35**	**46.83**	**47.66**	**47.65**	**47.55**	**0.052**	**0.185**	**0.008**	**0.058**

8.8.2 Type-b (Hazy) Model

H_d must be less than 50% of H_g to meet the requirements of this model, and sunlight accessibility must be between 7 and 9 h. The greatest power output of the hazy model, as shown in Table 8.5 is 99.45 W_p at 12:00 pm. When utilizing ANN, the values of MAPE and n-MAE were 0.067% and 0.072%, respectively; when using ANFIS, the values were 0.025% and 0.031%, respectively.

Figure 8.7 shows a graphical comparison between the measured and forecasted data for hazy weather. The total time duration considered was from 7 am to 17 pm. A maximum power of 99.45 W_p was achieved at 12:30 h. The sun's rays are hindered by haze, and temperature has a greater impact on solar radiation than time under sunny weather conditions. The sun irradiance hits 512.36 W/m^2 at its highest level.

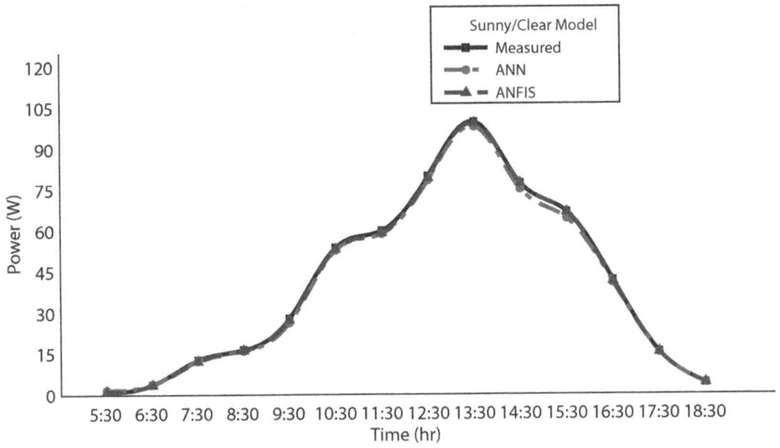

Figure 8.6 Forecasted and measured power for type-a model.

Table 8.5 ANN–ANFIS model for hazy weather.

Time (h)	Solar irradiance (W/m²)	Cell temp. (°C)	Power (W)			MAPE (%)		n-MAE (%)	
			Measured	ANFIS	ANN	ANFIS	ANN	ANFIS	ANN
7	11.83	6.48	1.41	1.41	1.87	0.071	0.125	0.084	0.101
8	46.67	6.59	7.25	7.26	8.36	0.035	0.106	0.023	0.112
9	92.37	11.36	14.78	14.75	13.89	0.025	0.145	0.010	0.125
10	146.36	14.52	20.59	20.60	21.71	0.018	0.112	0.002	0.104
11	284.78	20.36	46.35	46.33	47.82	0.016	0.086	0.026	0.026
12	512.36	35.78	99.45	99.43	98.75	0.026	0.025	0.048	0.046
13	470.85	40.23	88.12	88.10	89.35	0.043	0.031	0.052	0.037
14	330.56	35.89	60.23	60.21	61.38	0.021	0.035	0.063	0.046
15	182.59	27.26	31.46	31.47	29.03	0.017	0.026	0.041	0.051
16	54.36	17.59	8.59	8.52	7.57	0.014	0.037	0.013	0.063
17	12.85	11.87	2.47	2.45	1.39	0.015	0.085	0.014	0.071
Avg.	**178.79**	**18.99**	**31.72**	**31.71**	**31.76**	**0.025**	**0.067**	**0.031**	**0.072**

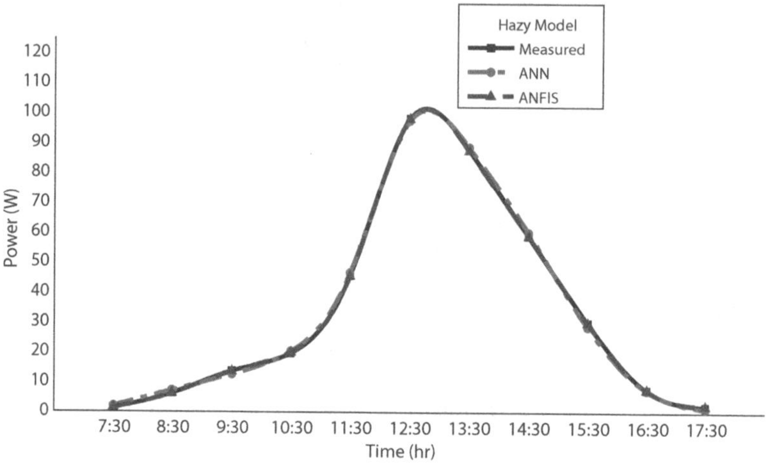

Figure 8.7 Forecasted and measured power for type-b model.

8.8.3 Type-c (Rainy) Model

To meet the requirements of this model, H_d must be less than 75% of H_g, and sunshine availability must be between 5 and 7 h. In this model, cloud cover and PV panels absorb a portion of the sun's energy. Table 8.6 illustrates the ANN–ANFIS model for rainy weather.

Power curve is biased around midday, with peak power output of 87.36 W_p around 15:00 h, according to Table 8.6. By using the ANN, MAPE and n-MAE achieved were 0.14% and 0.054%, respectively; however, with the ANFIS model, the values were significantly lower at 0.035% and 0.0014%, respectively. Figure 8.8 represents a graphical comparison between the measured and forecasted data for the rainy weather. The total time duration considered is from 7:00 am to 19:00 pm, and a maximum power of 87.36 W_p is achieved at 15:00 h.

The time range for rainy weather is from 7:00 am to 19:00 pm, and in these situations, solar radiation is partially absorbed by the PV and clouds. The solar irradiance reaches a maximum of 648.23 W/m^2. The absolute average error using the ANN was 0.14%, whereas the observed value using the ANFIS was 0.035%.

8.8.4 Type-d (Cloudy) Model

H_d must be greater than 75% of H_g and sunlight obtainability must be less than 5 h to meet the requirements of this model. Additionally, the data under consideration are averaged over a 24-hour period and Table 8.7

Table 8.6 ANN-ANFIS model for rainy weather.

Time (h)	Solar irradiance (W/m²)	Cell temp. (°C)	Power (W)			MAPE (%)		n-MAE (%)	
			Measured	ANFIS	ANN	ANFIS	ANN	ANFIS	ANN
7	29.36	27.85	5.12	5.12	4.98	0.023	0.12	0.0012	0.020
8	104.25	29.36	16.58	16.59	16.23	0.056	0.18	0.0036	0.042
9	185.32	32.85	25.12	25.12	25.36	0.064	0.16	0.0025	0.059
10	341.23	37.12	53.96	53.97	53.86	0.047	0.19	0.0048	0.063
11	426.85	41.59	58.02	58.02	57.23	0.036	0.14	0.0062	0.048
12	526.36	45.86	71.78	71.79	70.96	0.048	0.15	0.0037	0.069
13	504.29	45.47	67.36	68.37	67.12	0.031	0.14	0.0015	0.078
14	485.46	43.89	69.02	69.36	68.93	0.014	0.05	0.0039	0.035
15	648.23	50.28	87.36	87.36	87.25	0.019	0.17	0.0045	0.033
16	560.15	49.37	67.49	67.50	67.25	0.025	0.19	0.0026	0.020
17	330.78	40.58	39.12	39.13	39.02	0.037	0.12	0.0011	0.081
18	269.47	38.39	29.23	29.24	29.13	0.012	0.14	0.0023	0.070
19	87.30	32.78	8.96	8.97	8.95	0.019	0.03	0.0018	0.032
Avg.	**366.58**	**42.94**	**49.92**	**50.04**	**48.94**	**0.035**	**0.14**	**0.0014**	**0.054**

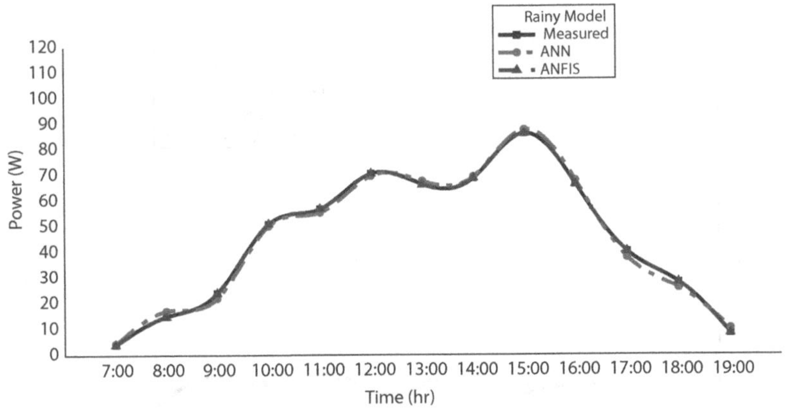

Figure 8.8 Forecasted and measured power for type-c model.

below shows the ANN–ANFIS model for type-d (cloudy) weather conditions.

Table 8.7 demonstrates that the power curve is at peak during mid-noon, giving peak power of 23.96 W_p at 10:00 a.m. The MAPE and n-MAE

Table 8.7 ANN–ANFIS model for cloudy weather.

Time (h)	Solar irradiance (W/m²)	Cell temp. (°C)	Power (W)			MAPE (%)		n-MAE (%)	
			Measured	ANFIS	ANN	ANFIS	ANN	ANFIS	ANN
7	14.92	10.32	2.36	2.36	2.57	0.023	0.225	0.005	0.052
8	100.02	11.58	19.36	19.36	17.68	0.034	0.124	0.007	0.046
9	125.61	16.49	17.48	17.48	20.37	0.058	0.169	0.002	0.036
10	115.23	19.36	23.96	23.96	22.49	0.069	0.145	0.006	0.047
11	102.56	16.78	16.28	16.29	17.92	0.047	0.173	0.008	0.060
12	140.25	17.25	23.56	23.58	24.64	0.023	0.182	0.003	0.038
13	112.58	17.69	19.78	19.78	21.69	0.069	0.126	0.002	0.061
14	116.39	17.25	21.72	21.72	21.68	0.074	0.216	0.001	0.034
15	46.38	14.23	7.36	7.37	8.21	0.081	0.147	0.008	0.043
16	25.23	15.36	4.02	4.03	4.51	0.026	0.189	0.0021	0.013
Avg.	**74.93**	**13.02**	**12.99**	**12.99**	**13.48**	**0.042**	**0.141**	**0.0036**	**0.035**

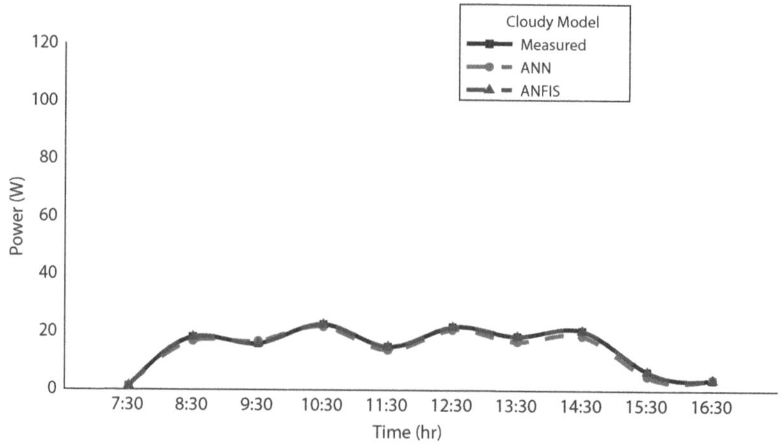

Figure 8.9 Forecasted and measured power for type-d model.

values for the ANN were 0.141% and 0.035%, respectively; however, they were significantly lower for the ANFIS model at 0.042% and 0.0036%, respectively.

The period considered in the fully foggy model is from 7:30 am to 16:30 pm, and around midday, the solar irradiation reaches its highest value of 140.25 W/m^2. Here, solar radiation is completely obscured by cloud cover, and the solar radiation is influenced by both time and temperature. When employing ANFIS instead of ANN, the absolute average error was measured to be 0.035% instead of 0.14%.

From Tables 8.4–8.7, the hazy model with MAPE of 0.0268% has outperformed other models for the composite climate zone, but type-a, type-c, and type-d models show significant differences with MAPEs of 0.0534%, 0.0610%, and 0.0493%, respectively. This could be due to the high relative humidity of the air, which varies from 64% to 81% during the wet season to 34%–60% during the dry season. Figures 8.6–8.9 demonstrate that the performance of the ANFIS model is superior. The hybrid approach used by ANFIS, which combines fuzzy logic and ANN, may be the cause of the higher convergence rate.

8.8.5 Comparative Analysis of the ANN–ANFIS Models with Fuzzy Logic Model

Additionally, the ANN–ANFIS model was compared with fuzzy models, and the results of this comparison are shown in Table 8.8, along with an evaluation of the model's performance using statistical evaluators for the composite climate.

Table 8.8 makes it clear that the suggested model performance is at par with any other model in global solar energy forecasting. For composite climatic conditions, the mean absolute percentage error observed by employing fuzzy models averaged 0.089%. Nevertheless, the use of intelligent models for forecasting global solar energy provides significantly better results. ANN model trained with back-propagation network has further lowered the MAPE to 0.011%. Finally, employing ANFIS models, MAPE further reduces to 0.0031%, and this finding reveals that ANN–ANFIS models are more accurate than other models.

The overall results show that ANN–ANFIS models reduce the amount of inaccuracy that can be acquired. Models based on ANFIS techniques have a certain mathematical structure, which makes them effective approximators. Additionally, the ANFIS-based model offers improved learning capabilities and lowers the convergence error for networks of comparable complexity.

Table 8.8 Comparative analysis of ANN–ANFIS model with fuzzy logic model.

Month	Solar irradiance (W/m^2)	Cell temp (°C)	Measured power (W)	ANN		ANFIS		Fuzzy	
				Forecasted power (W)	MAPE (%)	Forecasted power (W)	MAPE (%)	Forecasted power (W)	MAPE (%)
Jan	352.19	21.92	65.28	65.10	0.004	65.28	0.0003	65.04	0.09
Feb	449.87	29.27	90.23	90.21	0.016	90.23	0.0005	90.15	0.02
Mar	462.34	36.52	88.78	88.75	0.018	88.78	0.0022	88.62	0.12
Apr	485.12	39.36	76.36	76.35	0.026	76.36	0.0056	76.25	0.19
May	568.12	45.68	80.45	80.42	0.007	80.45	0.0003	80.38	0.03
Jun	539.64	43.58	69.78	69.75	0.003	69.78	0.0001	69.64	0.07
Jul	543.43	40.59	59.63	59.61	0.004	59.63	0.0024	59.54	0.04
Aug	422.78	39.78	60.12	60.10	0.030	60.12	0.0003	60.02	0.28
Sep	429.84	36.68	65.46	65.40	0.009	65.46	0.0011	65.32	0.13
Oct	475.82	38.89	89.17	89.15	0.002	89.17	0.0001	89.06	0.04
Nov	457.23	30.68	69.31	69.28	0.005	69.31	0.0011	69.15	0.01
Dec	468.45	24.58	77.45	77.41	0.009	77.45	0.0234	77.38	0.05
Avg.	**471.24**	**35.62**	**74.34**	**74.29**	**0.011**	**74.35**	**0.0031**	**74.21**	**0.089**

8.9 Conclusions

For PV power forecasting, this study established an ANN–ANFIS model with the aid of meteorological parameters for the composite climate zone. In this study, four models, i.e., sunny, hazy, rainy, and cloudy, were detailed and examined. After a comprehensive and thorough analysis, it was observed that the hazy model with an MAPE of 0.025% outperformed the sunny, rainy, and cloudy models with MAPE of 0.052%, 0.035%, and 0.042%, respectively, for composite climate conditions. Further, a comparative analysis of the proposed model was performed using a fuzzy-logic-based model. After a thorough analysis, it was observed that ANN–ANFIS models gave better performance results in comparison to fuzzy-logic-based models.

Acknowledgments

The authors appreciate the support of the National Institute of Solar Energy, India, for providing data for carrying out this present research work.

Conflict of Interest

All the authors declare that they have no conflict of interest.

ORCID

Gulnar Perveen https://orcid.org/0000-0001-6789-7455.

References

1. Duffie J. A. and Beckman W. A., *Solar Engineering of Thermal Processes*, 4th edition, Wiley, New York, April 2013.
2. Spencer J. W., *Fourier Series Representation of the Position of the Sun*, Search, vol. 2, no. 5, pp. 172–172, 1971.
3. https://mnre.gov.in/the-ministry/physical-progress, 2022.
4. IEA, Renewables 2018 – Market analysis and forecast from 2018 to 2023, https://www.iea.org/renewables, 2018.

5. Mayer M. J. and Grof G., Extensive comparison of physical models for photovoltaic power forecasting, *Appl. Energy*, 283, Article ID: 116239, pp. 1-18, 2021, DOI: 10.1016/j.apenergy.2020.116239.
6. Li G., Xie S., Wang B., Xin J., Li Y. and Du S., Photovoltaic power forecasting with a hybrid deep learning approach, *IEEE Access*, 8, pp. 175871-175880, 2020, DOI: 10.1109/ACCESS.2020.3025860.
7. Kejun W., Xiaoxia Q. and Hongda L., A comparison of day-ahead photovoltaic power forecasting models based on deep learning neural network, *Appl. Energy*, 251, Article ID: 113315, 2019. DOI:10.1016/j.apenergy.2019.113315.
8. Marcello A. F. B. L., Paulo C. M. C., Luis M. F. and Arthur P. S. B., Improving solar forecasting using deep learning and portfolio theory integration, *Energy*, 195, Article ID: 117016, 2020. DOI: 10.1016/j.energy.2020.117016.
9. Demirdelen T., Aksu I., Esenboga B., Aygul K., Ekinci F. and Bilgili M., A new method for generating short-term power forecasting based on artificial neural networks and optimization methods for solar photovoltaic power plants, *Solar Photovoltaic Power Plants: Advanced Control and Optimization Techniques*, pp. 165-189, 2019. DOI: 10.1007/978-981-13-6151-7_8.
10. Fara L., Alexandru D., Dan C. and Silvian F., Forecasting of energy production for photovoltaic systems based on ARIMA and ANN advanced models, *Int. J. Photoenergy*, 2021, Article ID: 6777488, pp. 1-19, 2021. DOI:10.1155/2021/6777488.
11. Massaoudi M., Abu-Rub H., Refaat S. S., Chihi I. and Oueslati F. S., Accurate smart-grid stability forecasting based on deep learning: point and interval estimation method, *IEEE Kansas Power and Energy Conference (KPEC)*, pp. 1-6, 2021. DOI: 10.1109/KPEC51835.2021.9446196.
12. Mehmet Y., Parameter extraction of photovoltaic cells and modules using Grey Wolf Optimizer with dimension learning-based hunting search strategy, *Energies*, 14(18), Article ID: 5735, 2021. DOI: 10.3390/en14185735.
13. Mohammed A. H., Abhinandan J. and Abu T., A novel Fast Mutable Duty (FMD) MPPT technique for solar PV system with reduced searching area, *J. Renew. Sustain. Energy*, 8, Article ID: 054703, 2016. DOI: 10.1063/1.4963314.
14. Naseem M., Mohammad A. H., Ahmad F. M., Ahmad N. K., Mohd A., Kumar J. D. and Iqbal A., Assessment of Meta-Heuristic and classical methods for GMPPT of PV system, *Trans. Electr. Electron. Mater.*, 22, pp. 217-234, 2021. DOI: 10.1007/s42341-021-00306-3.
15. Mohammad A. H., Abu T., Salman H., Saad Bin A. M. and Abhinandan J., Comparative assessment of maximum power point tracking procedures for photovoltaic systems, *Green Energy Environ.*, 2 (1), pp. 5-17, 2017. DOI: 10.1016/j.gee.2016.11.001.
16. Mohammad A. H., Abhinandan J., Abu T., and Atif I., Fast and precise global maximum power point tracking techniques for photovoltaic systems, *IET Renew. Power Gener.*, 13(14), pp. 2569-2579, 2019. DOI: 10.1049/iet-rpg.2019.0244.

17. Mohammad N., Mohammed A. H., Kumar J. D., Ahmad F. M., Anwar A., Syed M. A. and Akhtar A., A spider monkey optimization based global maximum power point tracking technique for photovoltaic systems, *International Conference on Emerging Frontiers in Electrical and Electronic Technologies*, INSPEC Accession Number: 21974961, India. DOI: 10.1109/CEFEET51821.2022.8947903.
18. Faiz A., Adil S., Farhad I. B., Shafiq A., Adam A. S. and Hafiz A., Parameter extraction of photovoltaic models using atomic orbital search algorithm on a decent basis for novel accurate RMSE calculation, *Energy Convers. Manage.*, 277, Article ID: 116613, 2023. DOI: 10.1016/j. enconmam.2022.116613.
19. Azka R., Soefian W., Aryani D. R., Jufri F. H. and Utomo A. R., Modelling of photovoltaic system power prediction based on environmental conditions using neural network single and multiple hidden layers, *2nd International Conference on Green Energy and Environment (ICoGEE 2020)*, IOP Conference: Earth and Environmental Science 599, 2020.
20. Ashraf I. and Chandra A., Artificial neural network based models for forecasting electricity generation of grid connected solar PV power plant, *Int. J. Glob. Energy Issues*, 21, 2004. DOI: 10.1504/ IJGEI.2004.004704.
21. Leary D. and Kubby J., Feature Selection and ANN Solar Power Prediction, *J. Renew. Energy*, 2017, Article ID: 2437387, pp. 1-7, 2017. DOI: 10.1155/2017/2437387.
22. Akhlaque A. A., Ahmad F. M., Rupendra K. P. and Hasmat M., Optimal sizing, control, and management strategies for hybrid renewable energy systems: A Comprehensive Review, *Energies*, 15, Article ID: 6249, 2022. DOI: 10.3390/en15176249.
23. Ahmad F. M., Mohammed A. H., Mohammed N. and Akhlaque A. K., Electricity demand modeling techniques for hybrid solar PV system, *Int. J. Emerg. Electr. Power Syst.*, 22(5), 2021. DOI: 10.1515/ijeeps-2021-0085.
24. Akhlaque A. A., Ahmand F. M., Qamar A. and Rupendra K. P., Energy demand modelling and ANN based forecasting using MATLAB/Simulink, *2021 International Conference on Control, Automation, Power and Signal Processing*, December 2021, Jabalpur, India. DOI: 10.1109/CAPS52117.2021.9730746.
25. Liu J., Linderholm H., Chen D., Zhou X., Flerchinger G. N., Yu Q., Du J., Wu D., Shen Y. and Yang Z., Changes in the relationship between solar radiation and sunshine duration in large cities of China, *Energy*, 82, pp. 589-600, 2015. DOI: 10.1016/j.energy.2015.01.068.
26. Fraihat H., Amneh A. A., Abdullah A., Bassam A., De F. and Paolo V., Solar radiation forecasting by Pearson Correlation using LSTM neural network and ANFIS method: Application in the West-Central Jordan, *Future Internet*, 14(3), Article ID: 79, pp. 1-24, 2022. DOI: 10.3390/fi14030079.
27. Ramadan A., Salah K., Ibrahim H. and Ahmed A., A novel intelligent ANFIS for the dynamic model of photovoltaic systems, *Mathematics*, 10, Article ID: 1286, pp. 1-14, 2022. DOI: 10.3390/math10081286.

28. Gafai N. B. and Belgore A. T., Feasibility of hybrid neuro-fuzzy (ANFIS) machine learning model with classical multi-linear regression (MLR) for the simulation of solar radiation: A case study Abuja, Nigeria, *Energy Res. J.*, 13(1), pp. 10-20, 2022. DOI: 10.3844/erjsp.2022.10.20.
29. Diaz N. P., Fernando A., Jose A. R., Yasuhiro M., Hussain A., Enrique J., Herrera L., Jess Fernando H., A. G., Ricardo A. and Enrique F. V., An ANFIS-based modeling comparison study for photovoltaic power at different geographical places in Mexico, *Energies*, 12(14), pp. 1-16, 2019. DOI: 10.3390/en12142662.
30. Muhammad N. A., Saad M., Hazlie M. and Noraisyah M. S., Review on forecasting of photovoltaic power generation based on machine learning and metaheuristic techniques, *IET Renew. Power Gener.*, 13(7), pp. 10091023, 2019. DOI: 10.1049/iet-rpg.2018.5649.
31. Munawar U. and Wang Z., A framework of using machine learning approaches for short-term solar power forecasting, *J. Electr. Eng. Technol.*, 15(2), pp. 561-569, 2020, DOI: 10.1007/s42835-020-00346-4.
32. Razagui A., Abdeladim K., Semaoui S., Hadj A. and Boulahchiche S., Forecasted power of a photovoltaic generator using numerical weather prediction and radiative transfer models coupled with a behavioral electrical model, *Energy Rep.*, 6(1), pp. 57-62, 2022. DOI: 10.1016/j.egyr.2019.08.018.
33. Srivastava S., Short term solar irradiation forecasting using ANFIS and simulated annealing ANFIS, *International Journal for Research in Applied Science and Engineering Technology*, 7, pp. 590-599, 2019. DOI:10.22214/ijraset.2019.11095.
34. Hossam F., Amneh A. A., Abdullah A., Bassam A., Roberto D. and Paolo V., Solar radiation forecasting by Pearson Correlation using ISTM neural network and ANFIS Method: Application in the West-Central Jordan, *Future Internet*, 14, Article ID: 79, pp. 1-23, 2019. DOI: 10.3390/fi14030079.
35. Bansal N. K. and Minke G., Climatic zones and rural housing in India, *Scientific Series of the International Bureau*, 35 (Kernforschungsanlage Julich, Julich, 1995).
36. Dolara A., Grimaccia F., Leva S., Mussetta M. and Ogliari E., Comparison of training approaches for photovoltaic forecasts by means of machine learning, *Appl. Sci.*, 8(2), Article ID: 228, 2018. DOI: 10.3390/app8020228.
37. Chel A. and Tiwari G. N., A case study of a typical 2.32 kWp stand-alone photovoltaic in a composite climate of New Delhi (India), *Appl. Energy*, 88, pp. 1415–1426, 2011.
38. Page J. K., The estimation of monthly mean values of daily total short wave radiation on vertical and inclined surfaces from sunshine records for latitudes 40°N–40°S, *Proceedings of UN Conference on New Sources of Energy*, Rome, Italy, 4, pp. 378–390, 1961.
39. Prescott J. A., Evaporation from the water surface in relation to solar radiation, *Trans. R. Soc. S. Aust.*, 64, pp. 114–118, 1940.

40. Riffonneau Y., Bacha S., Barruel F. and Ploix S., Optimal power flow management for grid-connected PV systems with Batteries, *IEEE Trans. Sustain. Energy*, 2(3), pp. 309-320, 2011, DOI: 10.1109/ TSTE.2011.2114901.
41. Hasmat M., Atif I., Puneet J., Sanjay A. and Farhad I. B., Metaheuristic and Evolutionary Computation: Algorithms and Applications, *Stud. Comput. Intell.*, 916, 2021.

9

Machine Learning Application for Solar PV Forecasting

Asif Khan[1]*, Mohd Khursheed[2], Jian Ping Li[3], Farhan Ahmad[1] and Ahmad Neyaz Khan[4]

[1]Integral University, Lucknow, Uttar Pradesh, India
[2]Department of Electrical Engineering, Integral University, Lucknow, Uttar Pradesh, India
[3]UESTC, Chengdu, Sichuan, China
[4]Department of Computer Science and Application, KL University, Vijayavada, India

Abstract
Owing to their intermittent nature, the integration of a substantial number of renewable energy sources (RESs), such as solar and wind, has an adverse impact on the stability and reliability of power systems. Solar PV power forecasting can be adopted to enhance system stability by providing estimated future power generation data to power system control engineers, as well as to optimize the dispatch of hydropower facilities. ML computational algorithms have demonstrated excellent performance in time sequence forecasting and can thus be used to anticipate power with weather factors as model inputs. This chapter describes the use of numerous ML computational algorithms for solar power forecasting in several solar parks in India. The performance of the ML algorithms in forecasting is compared to that of the Smart Persistence (SP) method, and the results reveals that the learning mode outperforms the other methods.

Keywords: Machine learning, solar PV, AI, prediction

9.1 Introduction

Flat-plate collectors are the most popular devices used to capture solar energy and transform it into thermal energy [1]. Solar ponds and saltwater

**Corresponding author*: asif05amu@gmail.com

Mohammed Aslam Husain, Md Waseem Ahmad, Farhad Ilahi Bakhsh, P. Sanjeevikumar and Hasmat Malik (eds.) Photovoltaic Systems Technology, (177–190) © 2024 Scrivener Publishing LLC

bodies intended to capture and store solar energy are other sources of thermal energy conversion. Solar energy can also be used to cook food in specially constructed solar ovens, which typically focus sunlight from a vast area to a central location. Solar energy can also be transformed directly into electricity via photovoltaic (PV) cells or solar cells [2]. Many gigawatt-scale PV power plants have been installed, and millions of installations are still ongoing, with solar energy accounting for half of all new-generation capacity in 2021. A PV system uses solar modules comprising several solar cells to generate electricity. PV systems can be installed on the ground, roof, wall, or water. A permanent mount or solar tracker can be used to track the sun's position to obtain the maximum power.

PV technology contributes to climate change mitigation by significantly emitting no carbon dioxide compared with conventional fuels. Solar PV has distinct advantages as an energy source, i.e., less pollution, no fuel cost, easily availability, and scalability in terms of power requirements. Furthermore, silicon is the backbone of the PV solar cell and is abundantly available in the Earth's crust [3]. Moreover, Concentrating Solar Power (CSP) can be used as a heat source in conventional power plants to generate electric energy [4]. CSP systems can also be used in agriculture and horticulture to maximize solar energy capture and increase plant productivity. Crop yields can be increased by using timed planting cycle schemes and adjusted row orientation [5], while sunshine is often seen as a plentiful resource, and the outliers highlight the necessity of solar energy in agriculture. Since the 1980s, engineers have been working on creating solar-powered car [1].

The main objective of applying machine learning techniques for solar PV forecasting is to improve the accuracy and reliability of predicting the amount of solar energy generated by photovoltaic (PV) systems. It is particularly important for power grid operators and utilities to maintain a balance between the generation and load demand of electric energy in real time to ensure grid stability. This chapter is divided into five sections. *Section 9.2* includes a literature review of the different models used for solar power forecasting. *Section 9.3*, include research on machine learning algorithms that can analyze large volumes of historical data on solar radiation, weather conditions, and PV system performance to identify patterns and make accurate predictions. By using these predictions to inform energy dispatch decisions, grid operators can better manage the variability in solar generation and reduce the need for costly backup power sources. *Section 9.4*, describe the experimental simulation and results based on solar PV. In addition to improving grid stability and reliability, accurate solar PV forecasting can benefit solar power plant operators by helping them optimize their energy production and maximize their revenue. Section 9.5, concludes by anticipating fluctuations in solar generation; plant operators can better plan maintenance schedules, avoid penalties for failing to meet contracted energy delivery requirements, and take advantage of market opportunities to sell excess energy.

9.2 Literature Review

The National Solar Mission is a major Indian government and a state government project to encourage environmentally green growth while solving India's energy security concerns [6]. Consequently, solar energy has enormous potential as a future energy source. However, there are several mathematical techniques for obtaining the maximum solar power under different cloudless conditions [1–3] explain three broadband models: Reference Evaluation of Solar Transmittance (REST), Reference Evaluation of Solar Transmittance, 2 band (REST 2), and Code for Physical Computation Radiation, 2 band (CPCR 2). To estimate solar energy, a regression model using single and multiple climatic characteristics as inputs has been proposed [7]. Clean energy has gained worldwide interest as the global economy has advanced and human environmental consciousness has grown. Solar energy has been employed in various industries as a clean renewable energy source, such as in power generation in solar photovoltaic systems. Several researchers have employed neural networks to forecast solar radiation as a function of weather variables [8]. Solar power systems are commonly employed owing to their low cost and high efficiency [9]. Compared to conventional energy sources such as oil, natural gas, and fossil fuels, it is regarded as one of the most likely Renewable Energy Sources (RES) because of its purity, quantity, and environmental friendliness [10]. Despite its benefits, the output power generated by a solar PV system change with changes in solar irradiance and operation temperature, particularly under rapidly changing partial shadowing conditions (PSC) owing to the nonlinear features of a photovoltaic (PV) cell. The complicated interaction between the power output and the PV input parameters leads to insufficient power extraction.

The operation of power and energy systems has changed dramatically [11]. These are mostly due to the increased penetration of renewable energy sources, which has been encouraged, for example, by the European Union (EU) under the H2020 Energy package [7]. Another notable development is the deregulation and restructuring of energy markets [12], which has resulted in the rise of a variety of new participants such as various types of aggregators [16]. The balance between consumption and generation is crucial when dealing with enormous amounts of Distributed Generation (DG) [13]. This equilibrium must be achieved both offline and online. In the online mode, balance and judgments are made in real-time. The offline mode makes decisions prior to an event [14]. The use of offline approaches necessitates the development of suitable resource-forecasting models (i.e., consumption and generation). Consequently, the quality of offline methods is inextricably related to the quality of forecasting methods [15]. Energy consumption is affected by a variety of factors (i.e., weather conditions, annual seasons, consumer behavior, etc.), making forecasting a complex and challenging task [16]. Regardless of its complexity, the electricity-consumption forecasting process should be reviewed in depth rather than in aggregate to ensure proper integration into demand response systems [17]. Demand response programs [18] enable small and medium-sized businesses to participate actively. One of the primary goals is to enable this participation autonomously without jeopardizing consumer comfort or the effective operation of energy services [19]. This autonomy can be quickly attained if the consumption forecast is thoroughly examined [20].

Yes, it is possible to address these concerns using data gathered from the solar power plants. The following approaches can be adopted:

1. Predicting power generation: By analyzing the power generation data and combining them with weather forecasts and historical weather data, it is possible to develop a predictive model that can estimate the expected electric energy generation for the next few days. This can be useful for grid management because it allows for better planning and utilization of the generated power.
2. Identifying the need for panel cleaning/maintenance: By analyzing sensor data and comparing it to historical data, it is possible to identify when panels require cleaning or maintenance. For example, a decrease in power generation that cannot be explained by changes in weather or other factors may indicate that the panels need cleaning. Additionally, if certain sensors indicate that the temperature or irradiance is outside the expected range, it may indicate that there is a problem with panels that require maintenance.
3. Identifying faulty or sub-optimally performing equipment: By analyzing the power generation data, it is possible to identify which inverters or lines of panels do not perform as

well as the others. By combining this information with sensor data, it is possible to identify the specific equipment that may be faulty or perform suboptimal. This information can be used to target maintenance efforts or identify equipment that may need to be replaced.

9.3 Research Methods and Materials

9.3.1 Dataset

In recent years, India has increased its emphasis on solar power generation, with the goal of producing 175 GW of renewable energy by 2022. These are some of the major data points regarding solar energy generation in India. Two solar power plants in India were studied for 34 days. Each plant contained numerous inverters, each of which was linked to multiple lines of solar panels. The power generation data were acquired at the inverter level, while sensor readings were obtained at the plant level using a single array of sensors.

The power generation data gathered at the inverter level can provide insights into the performance of each inverter, which can be useful for identifying any issues or inefficiencies in the system.

Overall, these data have the potential to provide valuable insights into solar power generation in India, and can help inform decisions related to plant design, operation, and maintenance. The feature datasheet is shown in Figure 9.1.

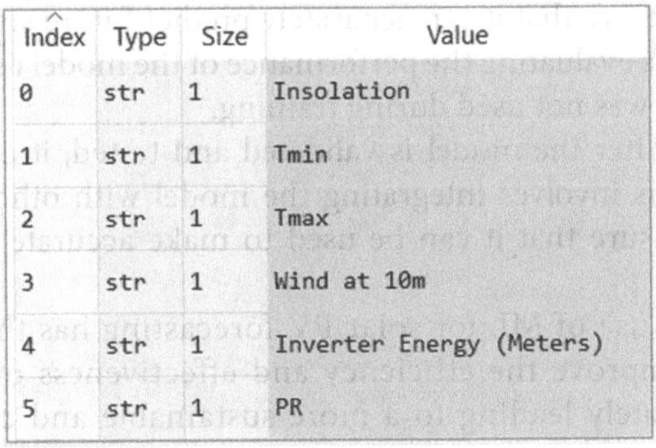

Index	Type	Size	Value
0	str	1	Insolation
1	str	1	Tmin
2	str	1	Tmax
3	str	1	Wind at 10m
4	str	1	Inverter Energy (Meters)
5	str	1	PR

Figure 9.1 Dataset features.

9.4 Proposed Work

Machine Learning (ML) has become an increasingly popular tool for solar PV forecasting due to its ability to learn complex patterns in large data sets. The following are some proposed steps for using ML for solar PV forecasting, as shown in Figure 9.2:

Data Collection: The first step involves collecting data from various sources, such as meteorological, PV plant, and weather data. The collected data should include historical data for a period of at least one year.
Data Cleaning and Preprocessing: The next step is to clean and preprocess the data to ensure that it is consistent and accurate. This involves the removal of outliers, missing values, and other errors.
Feature Selection: Once data are preprocessed, the next step is to select the most relevant features for the ML algorithm. This involves identifying the key variables that affect solar PV generation, such as solar irradiance, temperature, humidity, wind speed, and cloud cover.
Model Selection: Various ML algorithms can be used for solar PV forecasting, including neural networks, decision trees, and support vector machines. The choice of algorithm depends on the specific requirements of the project, such as the size of the dataset, complexity of the problem, and accuracy required.
Training the Model: Once model has been selected, it needs to be trained using historical data. The purpose of training is to educate the model to recognize patterns and correlations between the input and output variables (i.e., solar PV generation).
Validation and Testing: Once the model is trained, it must be validated and tested to ensure that it can accurately predict future solar PV generation. This entails evaluating the performance of the model using a subset of the dataset that was not used during training.
Deployment: After the model is validated and tested, it can be put into production. This involves integrating the model with other systems and processes to ensure that it can be used to make accurate predictions in real-time.

Overall, the use of ML for solar PV forecasting has the potential to significantly improve the efficiency and effectiveness of solar energy systems, ultimately leading to a more sustainable and cleaner energy future.

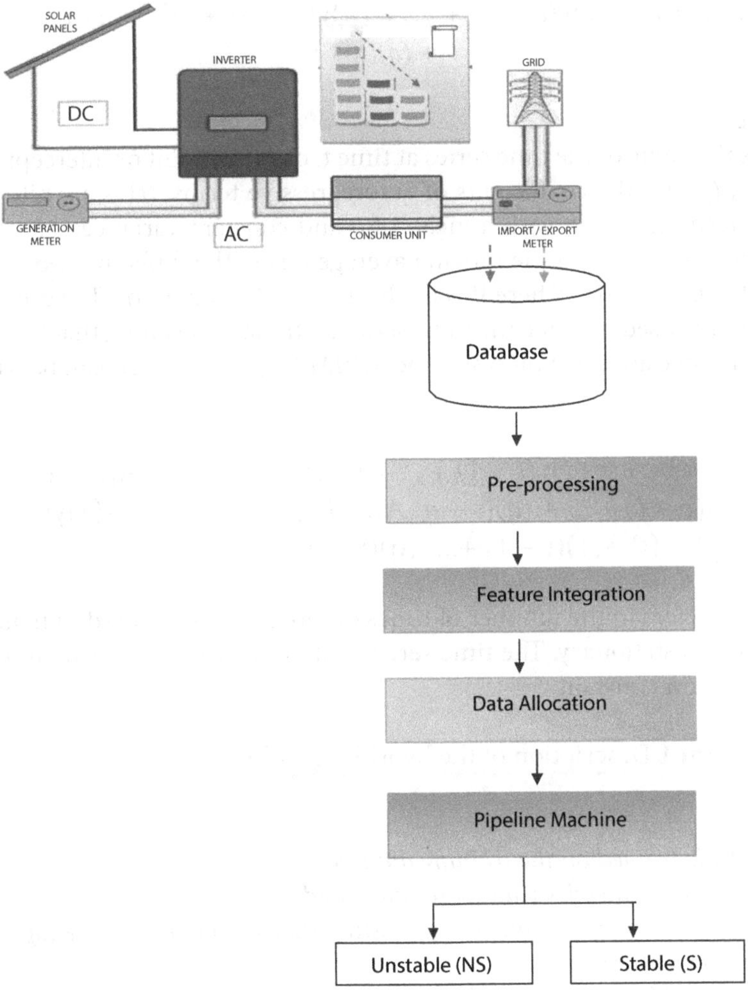

Figure 9.2 Model processing.

9.4.1 ARIMA Model

Autoregressive Integrated Moving Average (ARIMA) is a time-series forecasting method that simulates time-series behavior based on past values, trends, and seasonality. *ARIMA(p, d, q)* denotes the order of the autoregressive (AR) part, d represents the degree of differencing required to render the time-series stationary, and q represents the order of the moving average (MA) part. The ARIMA model can be mathematically represented as:

$$y(t) = c + (1)y(t-1) + \ldots + (p)y(t-p) + (t) + (1)(t-1) + \ldots + (q)(t-q)$$

where:
$y(t)$ is the value of the time series at time t. c is a constant or intercept term. $(1), \ldots, (p)$ are the coefficients of autoregressive terms. (t) is a white noise error term at time t, with a mean of 0 and constant variance[2]. $(1), \ldots, (q)$ are the coefficients of the moving average term. The difference operator is denoted by $(1 - B)^d$, where B is the backward shift operator. The difference operator is used to make the time-series stationary, meaning that the mean and variance are constant over time. $ARIMA(p, d, q)$ models can be written as

$AR(p)$: $y(t) = c + (1)y(t-1) + \ldots + (p)y(t-p) + (t)MA(q) : y(t) = c + (t) + (1)(t-1) + \ldots + (q)(t-q)$ $ARIMA(p, d, q) : y(t)\text{-}c\text{-}(1)y(t-1) \ldots (p)y(t\text{-}p) = (t) + (1)(t-1) + \ldots + (q)(t-q)$,

where d refers to the number of times differencing is required to make the time series stationary. The time series is stationary if both mean and variance remain constant.

Algorithm 1 Description of the hybrid algorithm

1: Start
2: *Model is trained on the training dataset.*
3: *Performance metrics values are calculated.*
4: *While casting predictions on the testing dataset, figure out the algorithm learning features*
5: *The model has now selected its feature by step1.*
6: *Feature set selection that render the lowest as well as highest scoring metrics*
7: End

9.5 Experimental Simulation, Result Analysis, Comparison, and Discussion

9.5.1 Data Reprocessing

Solar data reprocessing involves transforming and modifying existing solar data to improve their quality and usefulness for various applications in the

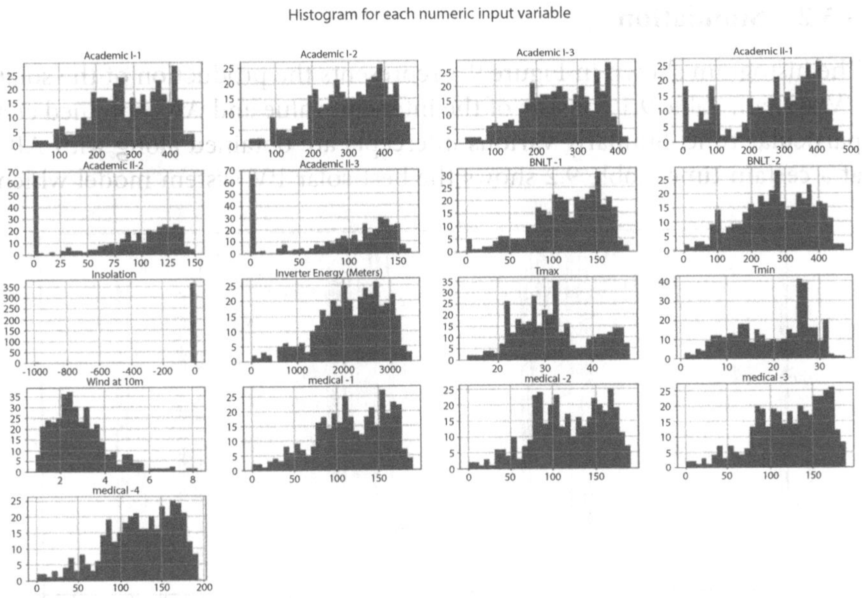

Figure 9.3 Hist decomposition of dataset.

solar energy industry. This can include tasks such as correcting instrument biases, aggregating data into higher temporal or spatial resolutions, and enhancing data with additional information such as weather data or information about the local power grid. Solar data reprocessing helps ensure the accuracy and reliability of the data for various purposes, including predicting solar energy production and supporting decision-making for solar energy system operators and investors. The Hist decomposition of the dataset is illustrated in Figure 9.3.

Algorithm 2 The proposed system algorithm.

1: *Preprocessing of genres solar power data using certain techniques.*
2: *Ensemble learning based feature selection.*
3: *Data partition using testing and training splits method.*
4: *Training the predictive ARIMA model on the training solar power data.*
5: *Validating the predictive ARIMA model using testing solar power data.*
6: *Computing the performance evaluation metrics of the model.*

9.5.2 Simulation

The histogram shown in Figure 9.4 represents the production of the solar PV system. Table 9.1 consists of the intercept value and AIC obtained at a particular time. Similarly various intercepts are obtained along with AIC at a certain time. Table 9.2 shows the best solar PV system model which

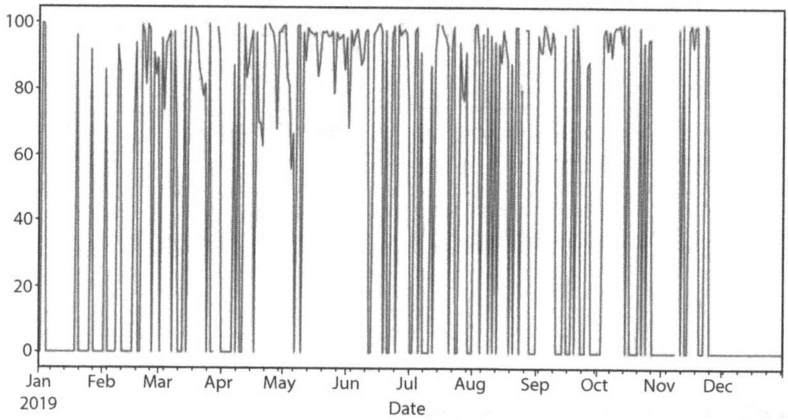

Figure 9.4 Ideal status of production of solar PV system.

Table 9.1 Model fit score.

Intercept	AIC	Time
(2,0,2)(0,0,0)[0]	3,287.998	0.83
(0,0,0)(0,0,0)[0]	3,362.962	0.45
(1,0,0)(0,0,0)[0]	3,361.348	1.08
(0,0,1)(0,0,0)[0]	3,351.662	2.07
(0,0,0)(0,0,0)[0]	3,365.084	1.57
(1,0,2)(0,0,0)[0]	3,280.564	1.90
(0,0,2)(0,0,0)[0]	3,283.544	1.90
(1,0,1)(0,0,0)[0]	3,320.919	1.76
(1,0,3)(0,0,0)[0]	3,283.560	2.01
(0,0,3)(0,0,0)[0]	3,287.600	2.61
(2,0,1)(0,0,0)[0]	3,305.175	1.85
(2,0,3)(0,0,0)[0]	3,290.967	1.96
(1,0,2)(0,0,0)[0]	3,281.142	1.43

Table 9.2 Accuracy outcome.

Name	Outcomes
Best mode	ARIMA(1,0,2)(0,0,0)[0] intercept
Total fit time	21.429 s
MSE	30.422626110684337
MAE	21.67315967281191

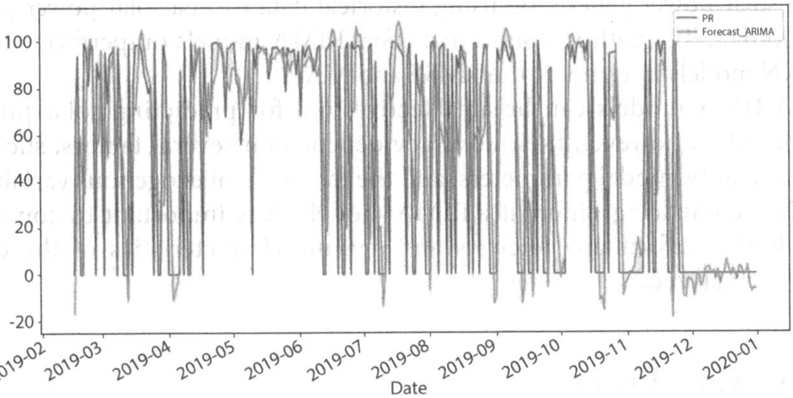

Figure 9.5 Forecasting.

consists of the model name, total time, MSE, and MAE. The solar PV power generation predicted by the proposed model is shown in Figure 9.5.

9.5.3 Comparison and Discussion

Solar power generation is a complex process that is influenced by various factors such as weather conditions, location, time of day, and seasonality. ARIMA models are often used for predicting solar power generation because they can capture the time-series nature of data and account for the impact of past values on future values.

In general, the accuracy of ARIMA models for predicting solar power generation depends on several factors, including the quality and quantity of historical data, choice of model parameters, and degree of seasonality and trends in the data.

When comparing different ARIMA models for solar power prediction, it is important to consider several metrics, such as the root mean squared error (RMSE), mean absolute error (MAE), mean absolute percentage error (MAPE), and coefficient of determination (R-squared).

Chen *et al.* (2018) compared the performance of ARIMA models in predicting solar power generation using historical data from a solar power plant in China. They found that ARIMA models with seasonality and trend components performed better than models without these components. Additionally, they found that including exogenous variables such as temperature and humidity improved the accuracy of the models.

Another study by Yassen and Al-Maqaleh (2019) compared the performance of ARIMA and artificial neural network (ANN) models in predicting solar power generation using historical data from a solar power plant in Jordan. The authors found that the ARIMA models outperformed the ANN models in terms of the RMSE and MAE.

ARIMA models can be an effective tool for predicting solar power generation; however, their accuracy depends on several factors, such as data quality, model parameters, and the inclusion of exogenous variables. When comparing different ARIMA models, it is important to consider multiple performance metrics and specific characteristics of the data being analyzed.

9.6 Conclusion

The machine learning ARIMA model worked effectively for solar PV forecasting. ARIMA models are particularly useful for capturing the time-series nature of solar power data and for accounting for the impact of past values on future values.

This study has shown that ARIMA models provide accurate predictions of solar power generation, especially when combined with seasonality, trend components, and exogenous variables such as weather conditions. However, the accuracy of ARIMA models depends on several factors, such as the quality and quantity of historical data, choice of model parameters, and degree of seasonality and trends in the data.

Overall, the models are useful for forecasting solar PV generation and can be applied in various settings, from large-scale solar power plants to residential rooftop solar panels. As solar energy continues to grow in popularity, the development of accurate forecasting models is becoming increasingly important for optimizing energy production and distribution.

References

1. J. Li, J. K. Ward, J. Tong, L. Collins, and G. Platt, "Machine learning for solar irradiance forecasting of photovoltaic system," *Renew. Energy*, vol. 90, pp. 542–553, 2016.
2. G. M. Tina, C. Ventura, S. Ferlito, and S. De Vito, "A state-of-art-review on machine-learning based methods for pv," *Appl. Sci.*, vol. 11, no. 16, p. 7550, 2021.
3. G. Narvaez, L. F. Giraldo, M. Bressan, and A. Pantoja, "Machine learning for site-adaptation and solar radiation forecasting," *Renew. Energy*, vol. 167, pp. 333–342, 2021.
4. L. Wang and J. Shi, "A comprehensive application of machine learning techniques for short-term solar radiation prediction," *Appl. Sci.*, vol. 11, no. 13, p. 5808, 2021.
5. A. I. Khalyasmaa, S. A. Eroshenko, V. A. Tashchilin, H. Ramachandran, T. Piepur Chakravarthi, and D. N. Butusov, "Industry experience of developing day-ahead photovoltaic plant forecasting system based on machine learning," *Remote Sens.*, vol. 12, no. 20, p. 3420, 2020.
6. R. Kabilan, V. Chandran, J. Yogapriya, A. Karthick, P. P. Gandhi, V. Mohanavel, R. Rahim, and S. Manoharan, "Short-term power prediction of building integrated photovoltaic (bipv) system based on machine learning algorithms," *Int. J. Photoenergy*, vol. 2021, pp. 1–11, 2021.
7. S. Theocharides, M. Theristis, G. Makrides, M. Kynigos, C. Spanias, and G. E. Georghiou, "Comparative analysis of machine learning models for day-ahead photovoltaic power production forecasting," *Energies*, vol. 14, no. 4, p. 1081, 2021.
8. K. Anuradha, D. Erlapally, G. Karuna, V. Srilakshmi, and K. Adilakshmi, "Analysis of solar power generation forecasting using machine learning techniques," in *E3S Web of Conferences*, vol. 309. EDP Sciences, 2021, p. 01163.
9. U. Munawar and Z. Wang, "A framework of using machine learning approaches for short-term solar power forecasting," *J. Electr. Eng. Technol.*, vol. 15, pp. 561–569, 2020.
10. M. N. Akhter, S. Mekhilef, H. Mokhlis, and N. Mohamed Shah, "Review on forecasting of photovoltaic power generation based on machine learning and metaheuristic techniques," *IET Renewable Power Gener.*, vol. 13, no. 7, pp. 1009–1023, 2019.
11. N. Sharma, P. Sharma, D. Irwin, and P. Shenoy, "Predicting solar generation from weather forecasts using machine learning," in *2011 IEEE international conference on smart grid communications (SmartGridComm)*. IEEE, 2011, pp. 528–533.
12. I. Jebli, F.-Z. Belouadha, M. I. Kabbaj, and A. Tilioua, "Prediction of solar energy guided by pearson correlation using machine learning," *Energy*, vol. 224, p. 120109, 2021.

13. A. K. Mittal, K. Mathur, and S. Mittal, "A review on forecasting the photovoltaic power using machine learning," in *J. Phys. Conf. Ser.*, vol. 2286, no. 1. IOP Publishing, 2022, p. 012010.
14. P. N. L. Mohamad Radzi, M. N. Akhter, S. Mekhilef, and N. Mohamed Shah, "Review on the application of photovoltaic forecasting using machine learning for very short-to long-term forecasting," *Sustainability*, vol. 15, no. 4, p. 2942, 2023.
15. M. Mohamed, A. Chandra, M. A. Abd, and B. Singh, "Application of machine learning for prediction of solar microgrid system," in *2020 IEEE International Conference on Power Electronics, Drives and Energy Systems (PEDES)*. IEEE, 2020, pp. 1–5.
16. E. Engel and N. Engel, "A review on machine learning applications for solar plants," *Sensors*, vol. 22, no. 23, p. 9060, 2022.
17. Z. Zulkifly, K. A. Baharin, and C. K. Gan, "Improved machine learning model selection techniques for solar energy forecasting applications," *Int. J. Renewable Energy Res. (IJRER)*, vol. 11, no. 1, pp. 308–319, 2021.
18. O. Bamisile, A. Oluwasanmi, C. Ejiyi, N. Yimen, S. Obiora, and Q. Huang, "Comparison of machine learning and deep learning algorithms for hourly global/diffuse solar radiation predictions," *Int. J. Energy Res.*, vol. 46, no. 8, pp. 10 052–10 073, 2022.
19. Y. Essam, A. N. Ahmed, R. Ramli, K.-W. Chau, M. S. Idris Ibrahim, M. Sherif, A. Sefelnasr, and A. El-Shafie, "Investigating photovoltaic solar power output forecasting using machine learning algorithms," *Eng. Appl. Comput. Fluid Mech.*, vol. 16, no. 1, pp. 2002–2034, 2022.
20. S. Theocharides, G. Makrides, A. Livera, M. Theristis, P. Kaimakis, and G. E. Georghiou, "Day-ahead photovoltaic power production forecasting methodology based on machine learning and statistical post-processing," *Appl. Energy*, vol. 268, p. 115023, 2020.

10

Techno-Economic Comparative Analysis of On-Ground and Floating PV Systems: A Case Study at Gangrel Dam, India

Satya Prakash Makhija[1], Pankaj Kumar Shrivastava[2]*, Prasanta Kumar Jena[3], Satya Prakash Dubey[4] and Pushpendra Singh[5]

[1]*Department of Electrical Engineering, Vindhya Institute of Technology and Science, Satna, India*
[2]*Department of Mechanical Engineering, Government Engineering College, Bhojpur (Bihar), India*
[3]*Department of Electrical and Electronics Engineering, BVRIT Hyderabad College of Engineering for Women, Hyderabad, Telangana, India*
[4]*ICFAI Universality, Raipur, Chhattisgarh, India*
[5]*Department of Electrical Engineering, Rajkiya Engineering College Banda, Uttar Pradesh, India*

Abstract
Solar energy is the most promising renewable energy source, because it is clean and freely available over its entire lifetime. Its primary flaw is that it requires a large amount of land, which restricts the use of this energy at a larger scale. To overcome this, floating PV systems can be installed over canals, ponds, water reservoirs, and dams to save land use and increase the power output by a large amount because of the cooling effect from water. Moreover, floating PV (FPV) can assist in maintaining water quality, reducing land use and water evaporation. In this chapter, a study was conducted to determine the feasibility potential of a 2,500 MW FPV system on Gangrel dam, Chhattisgarh, India, and it was compared with a nearby On-ground PV (OPV) system, using the System Advisory Model (SAM) software developed by NREL, USA. The FPV system has a positive net present value (NPV) of $60,386,996, whereas the OPV system has a negative value. The Internal rate of return (IRR) at the end of the project lifetime in FPV is 4.81%, which is greater than the value of 2.06% in the case of the OPV system. The annual

**Corresponding author*: psiitd@yahoo.com

Mohammed Aslam Husain, Md Waseem Ahmad, Farhad Ilahi Bakhsh, P. Sanjeevikumar and Hasmat Malik (eds.) Photovoltaic Systems Technology, (191–210) © 2024 Scrivener Publishing LLC

reduction in GHG emissions for floating PV was found to be 4,948.82 tCO_2/year. Calculations have also been performed for FPV systems to analyze the impact of the linked hydropower plant's enhanced hydropower output capacity.

Keywords: Floating, PV, dam, feasibility, environment

Description of Symbols/Abbreviations

Abbreviation/ symbol	Description	Abbreviation/ symbol	Description
SAM	System Advisor Model	M_{total}	No of required module
NREL	National Renewable Energy Laboratory	V_{max}	Max. MPPT voltage
NPV	Net Present value	V_{min}	Min. MPPT voltage
IRR	Internal Rate of Return	V_{mod}	Module voltage for Max. Power
GHG	Greenhouse Gas	$S_{parallel}$	Number of parallel Strings
CEA	Central Electricity Authority	C	Array Name Plate Capacity
CDIAC	Carbon Dioxide Information Analysis Centre	Inv_{total}	Total Number of Inverters
WEO	World Energy Outlook	$P_{max\text{-}mod}$	Maximum Module Power
MNRE	Ministry of New and Renewable Energy	R_{ratio}	Power Ratio of DC to AC

NASA	National Aeronautics and Space Agency	L	Length of module
LCOE	Levelized Cost of Energy	N	Operating Lifetime
MPPT	Maximum Power Point Tracking	E_t	Revenue from Electricity Sales
PPA	Power Purchase Agreement	C_t	Cost of Operation in Year t
FRL	Full Reservoir Level	r	Discount Rate
MCM	Million Cubic Metre	E_s	Electrical Energy Produced per year
C_o	Initial Capital Cost	β	Average Rate of Power Loss
W_{dc}	DC Power Rating of Module	Q	Water Discharge
G_t	Annual reduction in GHG emissions	H	Water Level/ Head in Reservoir
G	GHG emission (standard value) of each	η	Turbine Efficiency

10.1 Introduction

The relationship between the use of electric energy and the social development of a country is clear, suggesting that a country with having higher per capita energy consumption is in an advanced stage of development. India is developing country and as per reports of Government of India, per capita consumption of energy in India has increased from 631.4 kWh/yr to 1,075 kWh/yr since 2006 to 2016 (CEA 2018) [1]. Since the use of electric energy has become vital, it should be secure and sustainable, in addition to being economical, environmentally friendly, and socially acceptable. In January 2017, India installed a total 314,642.32 MW of power generation capacity of which 70% was from fossil fuels such as coal, diesel, oil, and nuclear sources. The use of fossil fuels causes greenhouse gas (GHG) emissions and their limited availability results in an increase in price. However, these fuels threaten the security, environmental cleanliness, and economic supply of energy. India has increased the yearly consumption of coal from 278 million tons to 545.9 million tons from 2006 to 2016. The Carbon Dioxide Information Analysis Centre (CDIAC), USA [2] report provides details of CO_2 emissions by country, showing China to have the highest 2,806,634 thousand metric tons of carbon emissions, followed by the USA and India, which have emissions of 1,432,855 and 610,411 thousand metric tons, respectively. The largest share of India's 30% increase in demand by 2014 comes from its 11% increase in global energy share (WEO 2017) [3]. Therefore, the installation of alternative and renewable energy systems is urgently required. Solar energy is the most promising renewable energy source because it is clean and freely available. The Ministry of New and Renewable Energy (MNRE) of India [4] has set a target of producing 175 GW of electrical power from renewable sources in India by 2022. China and India are leading countries in the rapid development of PV systems because of their vast solar energy potential. By 2040, when the percentage of all renewable energy sources in total power generation reaches 40%, solar energy will become the most important source of low-carbon energy (CEA 2018) [1]. One of the drawbacks of solar energy is that it has a large land demand, which is a significant problem in countries such as India, where the population and industries are continuously increasing. Therefore, the FPV system can be a viable option, and it can be used over canals, ponds, water reservoirs, and dams, saving land use and increasing the power output by a large amount because of the cooling effect from water. Moreover, FPV can aid in lowering water evaporation and preserving water quality. The land area saved can be used for forestation, and this will help decrease greenhouse gas emissions.

Sahu *et al.* [5] have given insight to the FPV technique, its current status, and various design options. They also described different aspects of the comparison between different PV installation techniques. Choi [6] compared the empirical data of an FPV system and an over-land PV system and found that the generation efficiency of floating PV systems increased by at least 11% compared to that of an over-land PV system. They also concluded that the temperature increase in PV modules of FPV is less than that in over-land PV systems because of the cooling effect of water. Gisbert *et al.* [7] have prepared the system's technical and financial viability by talking about the design elements of floating systems. Choi *et al.* [8] explained the optimal design plan of the tracking algorithm to implement a tracking scheme in PV panels to increase the power generation. Song *et al.* [9] offered to install FPV system on mine pit lakes and used system advisory model software to perform energy simulations. They calculated the NPV, payback period, and some financial and scientific parameters and compared them with those of the overland system. In comparison to on-land PV systems set in place after deforestation, it also demonstrated a halving of the GHG emissions. Kumar *et al.* [10] analyzed different kinds of power losses and efficiency ratio of a 10MW solar power plant and compared the result with PVSyst and PV-GIS software. Khan *et al.* [11] have analyzed the process parameters of hybrid PV-Wind-DG-Battery system for different storage technologies and locations. They found that lead-acid-based technology was the most suitable among the various other considerations. Liu *et al.* [12] have done the performance and cost analysis of a PV system for a rural household under various ways which contain fixed-tilt (FT), horizontal axis (HM), vertical axis (HV), and dual-axis tracker (DA). Kumar *et al.* [13] designed a 4 kWp rooftop solar PV system for a residential building in Motihari, Bihar, India to ensure a continuous supply of electricity. Pandey *et al.* [14] have presented the financial viability analysis for PV generation at REC, Ambedkar Nagar, UP, India

India is gifted with large-area water reservoirs, which are associated with hydropower generating plants and can be utilized for the installation of FPV systems. In the present study, a novel work was performed that presents a comparative analysis of FPV and OPV systems based on various technical, economic, and environmental aspects. In the case of FPV systems, estimates have also been made to assess the impact of the enhanced output capacity of connected hydropower plants. In this chapter, a detailed analysis of the feasibility of FPV systems on the Gangrel dam located in the Chhattisgarh state of India is presented. The reservoir has a surface area of 95 km^2, in which a 50 km^2 surface has the minimum water availability throughout the year, over which the FPV can be installed. The solar radiation over the area has an annual average radiation of 5.06 $kWh/m^2/day$. System Advisor Model (SAM) software was used to evaluate various financial and technical parameters of the system. A 2,500 MW solar system has been proposed for installation over the surface, and its generation throughout the year has been evaluated. Financial parameters, such as NPV, IRR, payback period, and technical parameters of generation and losses, were evaluated using SAM. The land area saved can be used for planting trees, which will reduce greenhouse gas emissions and shading by solar panels will improve water quality by preventing photosynthesis and algae growth.

The present chapter has been organized to include *Section 10.1* as an introduction to the field of floating PV systems. This is followed by project site assessment for various parameters in *Section 10.2* and the design of on-ground and FPV systems in *Section 10.3*. The simulation of the two

Table 10.1 Position of water in Gangrel dam.

Sl. no.	Year	Date	Water level		
			Live capacity (MCM)	Water in dam (MCM)	%age of live capacity
1	December 2016	15.12.2006	766.90	762.18	99.38
2	January 2017	16.01.2007	766.90	725.42	94.59
3	February 2017	15.02.2008	766.90	641.25	83.62
4	March 2017	15.03.2009	766.90	546.37	71.24
5	April 2017	01.04.2017	766.90	457.39	57.34
6	May 2017	01.05.2017	766.90	209.59	27.33
7	June 2017	16.06.2017	766.90	89.65	11.69
8	July 2017	06.07.2017	766.90	130.51	17.02
9	August 2017	06.08.2017	766.90	255.66	33.34
10	September 2017	06.09.2017	766.90	314.94	41.07
11	October 2017	06.10.2017	767.00	252.60	32.93
12	November 2017	06.11.2017	767.00	301.09	39.26

systems, results, and analysis are in *Section 10.4* and finally, concluding remarks are presented in *Section 10.5.*

10.2 Project Site Assessment for Various Parameters

The Gangrel dam is located in the Dhamtari district, Chhattisgarh, India, and is built across the Mahanadi River. It has a length of 1,830 m and a surface area of 95 km^2. A small part of the dam was used to produce electric energy from a hydroelectric power plant with a capacity of 10 MW. The position of water in the dam reservoir is given in Table 10.1 from December 2016 to November 2017 as obtained from the Ministry of Water Resources, Government of Chhattisgarh [15]. In most months, the water level was found to be less than the live capacity. It has been found that a minimum water level is found in the month of June leading to less amount of power generation in the hydropower plant. Figure 10.1 depicts a satellite view of the project location where the dam area is located [16]. The meteorological characteristics of the site are listed in Table 10.2 (accessed at the location). A minimum surface area of 50 km^2 is available throughout the year over which photovoltaic system has been proposed to be installed.

The solar radiation profile has been accessed from NASA meteorological department for the project location and is provided in Figure 10.2. It gives the graphical representation of assessed monthly solar radiation. It is observed that in the month of April and May, maximum radiation is available at Gangrel dam [17]. Figure 10.3 shows the average monthly daily temperature accessed from NASA website at the latitude and longitude of Gangrel dam.

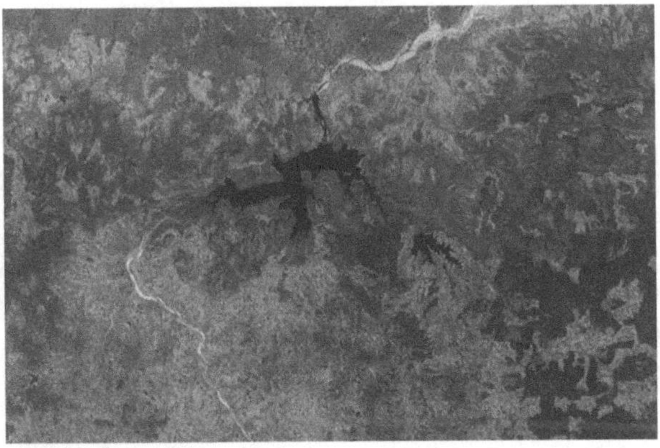

Figure 10.1 Satellite view of Gangrel dam.

Table 10.2 Meteorological parameters.

Item	Value
Latitude (deg)	20.616678
Longitude (deg)	81.564431
Monthly average direct normal radiation (kWh/m²/day)	5.32
Surface area (km sq.)	95
Height (feet)	30.5
Length (meter)	1,830

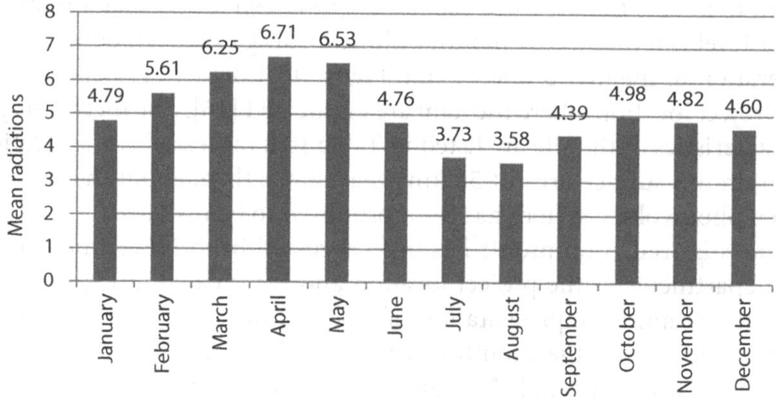

Figure 10.2 Monthly daily radiations (kWh/m²/day).

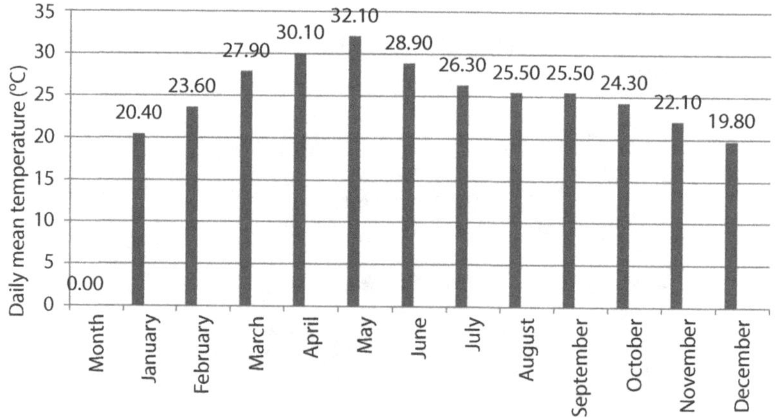

Figure 10.3 Daily mean temperature at the project location.

10.3 Design of On-Ground and Floating PV Systems

10.3.1 On-Ground Photovoltaic System

In this case a fixed orientation of panels has been considered and it has been planned to face the solar panels in the south direction since India lies in northern hemisphere. The system is installed on the ground surface which is near by the dam area. A 2500MW of PV system has been proposed to be installed over the 50square km surface area. The following parameters have been considered in designing the photovoltaic system [9]. The number of required modules to be used is obtained from

$$M_{total} = [\frac{Vmax + Vmin}{2}]/V_{mod} \tag{10.1}$$

$$S_{parallel} = [(C * 1000W/KW)/P_{max}\}/M_{total} \tag{10.2}$$

where V_{max} and V_{min} are the maximum and minimum MPPT voltage in the inverter. V_{mod} represents maximum power voltage of the module. $S_{parallel}$ is the no of parallel strings and P_{max} is the maximum module power.

Total number of inverters required are given by

$$(\mathrm{Inv}_{total}) = (\mathrm{M}_{total} * \mathrm{S}_{parallel} * \mathrm{P}_{max-mod}) / (\mathrm{R}_{ratio} * \mathrm{P}_{inv-max}) \tag{10.3}$$

Here, $P_{max-mod}$ is the maximum module power, R_{ratio} represents the dc to ac ratio and has been given value 1.0 $P_{inv-max}$ is the maximum ac power of the inverter.

The spacing between arrays has been calculated as

$$X_{spacing} = \mathrm{L} * \{\cos(\mathrm{tilt}) + \sin(\mathrm{tilt}) * \tan(\mathrm{Lat} + 23.5)\} \tag{10.4}$$

Here, tilt is the tilt angle in degrees and L is the length of the module.

The needed total area of the solar site is determined as follows, once the array spacing and tilt angle of each string are known

$$\text{Required area} = (\mathrm{M}_{total} * \mathrm{M}_{width}) * \mathrm{X}_1 * \mathrm{S}_{parallel} \tag{10.5}$$

The NPV of a project is calculated by the following equation (Short, Packey, and Holt 1995)

$$NPV = \sum_{t=1}^{N} \frac{E_t - C_t}{(1+r)^t} - C_0 \tag{10.6}$$

where NPV is the net present value, N is the system operating period (year), E_t is the electricity sales revenue in year t, C_t is the operating cost in year t, r is the discount rate, and C_0 is the initial cost.

The Internal Rate of Return (IRR) is the rate that sets the NPV equal to zero. The levelized cost of energy (LCOE) is another parameter to be analyzed to determine the performance of PV systems. It is determined by dividing the total annualized cost of the system by the amount of electricity produced each year [18].

The lifetime of the proposed system is considered to be 25 years. The total installed cost of the solar panel along with the inverter cost has been considered to be \$0.76/W_{dc} [19] and land cost at the project location is \$3,000/acre per local land cost market. Thus, the total installation cost was calculated as \$0.77/W_{dc}. The operation and maintenance costs are \$15/kW-yr, with an escalation rate of 0.27% per year. Financial parameters, such as taxes and interest rates, are considered per state and central government norms. The PPA price of electricity generated is \$0.044/kWh [19], with a price escalation of 2% per annum.

10.3.2 Floating PV System

SAM lacks the ability to model FPV systems, whose installation designs differ from those of on-ground PV systems, and are more expensive. The technical and panel cost criteria considered have the same values as those in the on-ground system. Owing to the need for frames and supporting systems to make it float, the installation cost is 1.2 times greater than that of the on-ground system [9]. Figure 10.4 depicts a typical floating construction [5]. Because an FPV system is spread over the surface of the water, land costs are eliminated. In an experimental study, it was discovered that FPV systems produce 11% more power than on-ground systems. [7]. Hence, accounting for the above parameters, to compare the financial parameters of the OPV system with those of the FPV system, the system costs in the FPV were considered to be \$0.80/W_{dc} which was \$0.76/W_{dc}. The FPV system was simulated for the same 2,500 MW capacity and performance parameters using SAM.

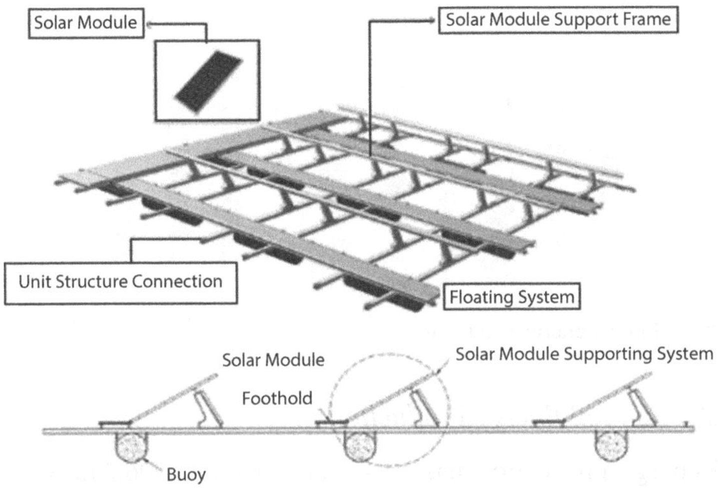

Figure 10.4 Floating PV supporting structure.

10.4 Simulation, Results and Analysis

The SAM software has been used to evaluate financial and technical performance of the two systems, i.e., OPV and FPV systems. In the present work, these two 2,500 MW solar

Since SAM does not have facility to design FPV systems, initially OPV system has been modeled and simulated to get its performance parameters. Subsequently, by changing the capital cost and eliminating land cost, the results of the FPV system were obtained. The performance of the two systems is discussed in the following sections.

10.4.1 On-Ground PV System

10.4.1.1 Monthly Energy Production

When solar energy is incident on the solar panels of the OPV system, the modules convert irradiance into useful power. The power output varies with solar irradiance and temperature. The results of monthly energy production are shown in Figure 10.5. It can be observed that the energy output was maximum in March.

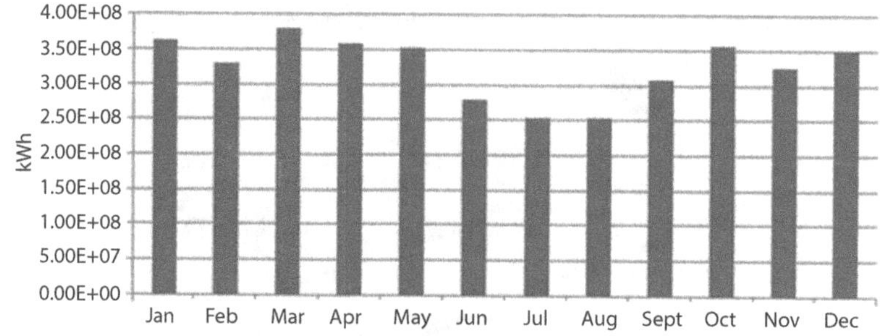

Figure 10.5 Monthly energy production.

10.4.1.2 Annual Energy Production

Annual energy production has been found to be 3,906,246 MWh. The annual energy production was calculated over a lifetime of 25 years. Because the degradation has been neglected, the output seems to be constant over the years. Figure 10.6 represents the annual energy production at the project location.

10.4.1.3 Loss Diagram

When solar energy occurs on the solar panels, it converts heat energy into DC electrical energy. The conversion efficiency of the solar panel used is 16.89%. The DC energy is converted into AC by passing it through an inverter with an efficiency of 97.59%. During the process, different types of component losses occur, such as diode loss and dc wiring loss. Figure 10.7 shows a loss diagram of the proposed system.

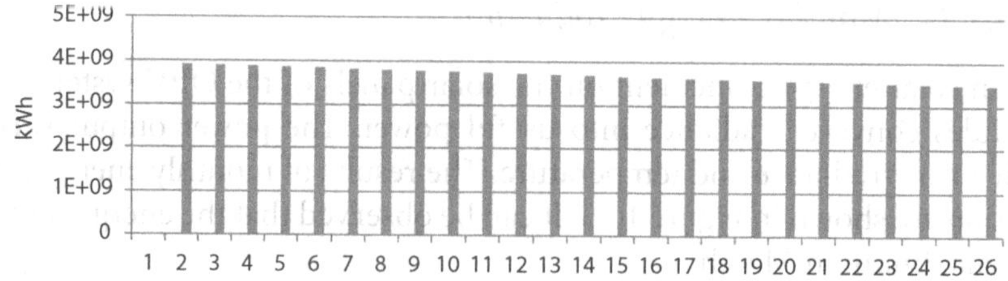

Figure 10.6 Annual energy production.

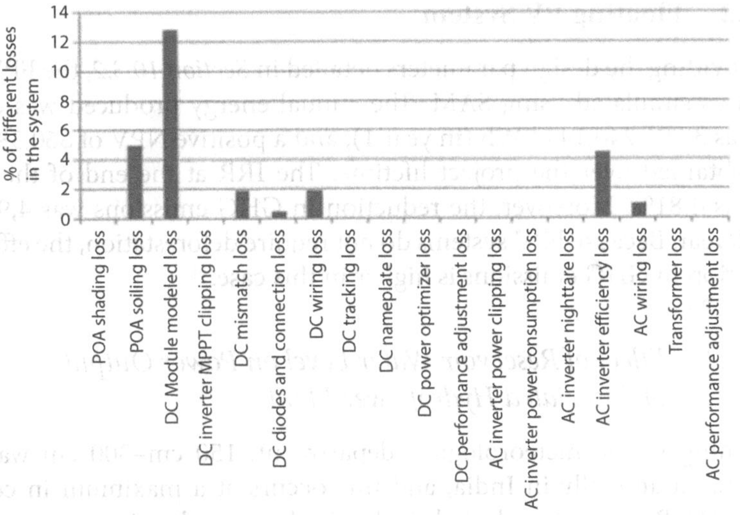

Figure 10.7 Loss diagram.

10.4.1.4 Analysis of Greenhouse Gas Emission

The reduction in greenhouse gas emissions is the actual amount of greenhouse gas generated when a fossil fuel system is used to generate the same amount of electricity produced by a renewable energy system (Song and Choi 2016). The amount of greenhouse gas is calculated by

$$G_t = E_s * 10^{-3} * G * (1 + \beta) \tag{10.7}$$

G_t represents the annual reduction in GHG emissions (tCO2/year) and E_s represents the annual power output from the PV system (MWh/year). β is the average loss rate of electricity transmission and distribution, which is 0.23 in the case of the project location and is the standard amount of GHG emissions for each country (tCO2/MWh). The value of β depends on the temperature and humidity profiles of the system. Thus, the amount of reduction in GHG emissions is determined by both clean energy production and loss rate β. The emission factor was assumed to be 1.030 kg CO_2/kWh [20]. In the proposed model, to produce 3,906,246,144 kWh of electrical energy, the reduction in GHG emissions was calculated to be 4,948.82 tCO_2/year.

10.4.2 Floating PV System

By providing the design parameters detailed in *Section 10.3.2*, the FPV system was simulated using SAM. The annual energy produced was calculated as 3,906,246,144 kWh (in year 1), and a positive NPV of $60,386,996 was obtained over the project lifetime. The IRR at the end of the project was 4.81%. Moreover, the reduction in GHG emissions was 4,948.82 tCO2/year. Because FPV systems do not require deforestation, the effective reduction in GHG emissions is higher in this case.

10.4.2.1 Effect of Reservoir Water Level on Power Output of Associated Hydropower Plant

According to the meteorological department, 150 cm–300 cm water is evaporated annually in India, and this occurs at a maximum in central India [21]. Because the selected site lies in the central region, water evaporation is one of the main concerns in this open-area dam. A research study carried out in Australia suggests that FPV systems can reduce water evaporation by approximately 50% in man-made facilities such as dam reservoirs [5]. The average reduction in water level due to evaporation was calculated for the considered time period data, as shown in Table 10.1, and was calculated to be 195.27 MCM. FPV systems may increase the water level by 50%; thus Table 10.3 shows the water level of the reservoir covered with floating systems. An increased water level will certainly increase the power generation capacity of the associated hydropower plant, as described by Eq. (10.8) [19].

$$P = (\eta\ QH) / (1.181 * 10^4) \tag{10.8}$$

where, P is power (MW), η is turbine efficiency (%), Q is discharge (ft/s), and H is water level/head (ft).

Equation (10.8) suggests that the power output of a hydropower plant is directly proportional to its reservoir water level (H). Based on this, the percentage increase in the water level and hence power generating capacity of the hydroplant is shown in Table 10.4. An increase in power generation due to the floating PV installment was calculated as compared to the OPV instalment over the considered 12 months and the average increase in power generation in the hydropower plant was estimated to be more than 21% due to the increase in water level due to less evaporation.

Table 10.3 Position of water in Gangrel dam.

		Water level				
Sl. no.	**Year**	**Date**	**Open reservoir (MCM)**	**%age of live capacity at F.R.L**	**Reservoir covered with floating PV modules**	**%age of live capacity at F.R.L**
1	December 2016	15.12.2006	762.18	99.38	957.45	100
2	January 2017	16.01.2007	725.42	94.59	920.69	100
3	February 2017	15.02.2008	641.25	83.62	836.52	100
4	March 2017	15.03.2009	546.37	71.24	741.64	96.7
5	April 2017	01.04.2017	457.39	57.34	652.66	85.1
6	May 2017	01.05.2017	209.59	27.33	404.86	52.79
7	June 2017	16.06.2017	89.65	11.69	284.92	37.15
8	July 2017	06.07.2017	130.51	17.02	325.78	42.45
9	August 2017	06.08.2017	255.66	33.34	450.93	58.79
10	September 2017	06.09.2017	314.94	41.07	510.21	66.52
11	October 2017	06.10.2017	252.60	32.93	447.87	58.39
12	November 2017	06.11.2017	301.09	39.26	496.36	64.71

Table 10.4 Percentage in water level and power generation capacity of hydroplant after installation of floating PV system.

Sl. no.	Month, year	% increase in water level and power generating capacity*
1	December 2016	0.62
2	January 2017	5.41
3	February 2017	16.38
4	March 2017	25.46
5	April 2017	27.76
6	May 2017	25.49
7	June 2017	25.46
8	July 2017	25.43
9	August 2017	25.45
10	September 2017	25.45
11	October 2017	25.46
12	November 2017	31.75

10.4.2.2 Effect on PV System Structure Material, Flora–Fauna of Water and Other Activities

The air in the surroundings of the reservoir has a higher moisture content. This will certainly result in the increased corrosion of the metallic structure and components, which can reduce the life of the system. FPV systems do not allow sunlight to penetrate the water body and thus affect the growth of aquatic animals and seaweeds [5]. FPV systems may also affect water transportation and fishing activities, depending on the location of the site.

10.4.3 Comparative Analysis Between On-Ground PV System and Floating PV System

After simulating the OPV and FPV systems for over 25 years at the project location, it was observed that the FPV system is an economically viable solution. Table 10.5 represents the calculated financial parameters for the two systems.

Table 10.5 Comparison between on-ground PV and floating PV parameters.

	On-ground PV system	**Floating PV system**
Metric	Value	Value
Annual energy (year 1)	3,906,246,144 kWh	3,906,246,144 kWh
Levelized PPA price (real)	3.14 ¢/kWh	4.31 ¢/kWh
Levelized COE (real)	5.44 ¢/kWh	4.25 ¢/kWh
Net present value	$1,208,442,880	$60,386,996
IRR at end of project	2.06%	4.81%
Net capital cost	$1,967,424,384	$2,030,728,832
Minimum DSCR	1.58	2.28

10.4.3.1 Comparison Based on Other Parameters

Apart from financial parameters, there are other parameters that are tabulated in Table 10.6. The effects of the parameters have been discussed in the results section of each PV system.

10.5 Conclusion

The present study has designed of two 2,500 MW solar power plants, one on-ground PV (OPV) system, and another floating PV system (FTV). Both the OPV and FPV systems were simulated using the System Advisor Model (SAM) after feeding various technical, financial, and meteorological data. The performances of the two systems were evaluated by analyzing various parameters. The comparative analysis revealed that the FPV system has a positive NPV of $60,386,996 whereas the OPV system is found to have negative NPV of $1,208,442,880. A positive NPV makes an FPV system financially feasible. Other additional benefits of installing FPV systems include higher power generation for the same panel area, less maintenance, land area savings, and reductions in greenhouse gas emissions. Because of less water evaporation, FPV systems are also advantageous with respect to the higher generation capacity of associated hydropower plants, water saving, and retention of quality. The present study will certainly benefit government and private agencies in making decisions for their future PV system projects.

Table 10.6 Comparison based on miscellaneous parameters.

Parameter	On-ground PV system	Floating PV system
Power generating efficiency	Lower efficiency due to higher surrounding temperature	Higher efficiency due to lower surrounding temperature in the vicinity of water
Power generating capacity	Lower efficiency leads to lower generation capacity	Higher efficiency results more power production capacity
Maintenance with respect to cleaning	More maintenance due to higher dust accumulation over module surface	Lower maintenance due to very low dusty environment over water surface
Maintenance of PV system structure	Lower maintenance due to less complexity of system structure	Higher maintenance due to complex supporting structure
Control strategy for ease of maintenance	It is very easy due to their accessibility without any support platform on ground	Separate supporting pathways are prepared between module strings for ease of maintenance.
Environmental effects	It is a clean energy system	It is a cleaner system since it avoids deforestation.
Degradation of PV system structure material	Very less degradation	More degradation due to higher moisture in surroundings as it floats over water
Flora–fauna effect	Not applicable in this case	It may affect the life and growth of aquatic animals and seaweed
Water transportation and fishing activities	Not Applicable in this case	These activities may be affected depending on the location of reservoir.

References

1. CEA. 2018. "Executive Summary of Power Sector." GOVERMENT OF INDIA, Last Modified January 30,2018, accessed March 5,2018. http://www.cea.nic.in/monthlyexesummary.html.
2. CDIAC. 2018. "Fossil-Fuel CO2 Emissions." CDIAC, Last Modified September 26,2017, accessed March 11,2018. http://cdiac.ess-dive.lbl.gov/trends/emis/top2014.tot.
3. "World Energy Outlook 2017." International Energy Agency, accessed 02 Feb. https://www.iea.org/weo2017.
4. MNRE. 2018. "Minister of New and Renewable Energy." Goverment of India, Last Modified May 29,2018, accessed March 5,2018. https://mnre.gov.in/#.
5. Sahu, Alok, Neha Yadav, and K Sudhakar. 2016. "Floating photovoltaic power plant: A review." *Renewable Sustainable Energy Rev.* 66:815-824.
6. Choi, Young-Kwan. 2014. "A study on power generation analysis of floating PV system considering environmental impact." *Republic of Korea.*
7. Ferrer-Gisbert, Carlos, José J Ferrán-Gozálvez, Miguel Redón-Santafé, Pablo Ferrer-Gisbert, Francisco J Sánchez-Romero, and Juan Bautista Torregrosa-Soler. 2013. "A new photovoltaic floating cover system for water reservoirs." *Renewable Energy* 60:63-70.
8. Choi, Young-Kwan, Nam-Hyung Lee, An-Kyu Lee, and Kern-Joong Kim. 2014. "A study on major design elements of tracking-type floating photovoltaic systems." *Int. J. Smart Grid Clean Energy* 3(1):70-74.
9. Song, Jinyoung, and Yosoon Choi. 2016. "Analysis of the potential for use of floating photovoltaic systems on mine pit lakes: Case study at the ssangyong open-pit limestone mine in Korea." *Energies* 9 (2):102.
10. Kumar, B Shiva, and K Sudhakar. 2015. "Performance evaluation of 10 MW grid connected solar photovoltaic power plant in India." *Energy Rep.* 1:184-192.
11. Khan, F.A., Pal, N., Saeed, S.H. and Yadav, A., 2022. Techno-economic and feasibility assessment of standalone solar Photovoltaic/Wind hybrid energy system for various storage techniques and different rural locations in India. *Energy Convers. Manage., 270*, p.116217.
12. Liu, X., Tan, Q., Niu, Y. and Babaei, R., 2022. Techno-economic analysis of solar tracker-based hybrid energy systems in a rural residential building: A case study in South Africa. *Int. J. Green Energy*, pp.1-20.
13. Kumar, A., Andleeb, M. and Bakhsh, F.I., 2021, March. Design and analysis of solar PV rooftop in Motihari. In *J. Phys.: Conf. Ser.* (Vol. 1817, No. 1, p. 012019). IOP Publishing.
14. Pandey, A., Iqbal, A., Mishra, A.K. and Husain, M.A., 2019, February. Economic Feasibility Analysis of Solar PV Generation at REC Ambedkar Nagar. In *2019 IEEE International Conference on Electrical, Computer and Communication Technologies (ICECCT)* (pp. 1-6). IEEE.

15. CGWRD. 2017. "Chhattisgarh Water Resources Department." Goverment of Chhattisgarh, accessed March 11,2018. http://www.cgwrd.in/docs-industry/cat_view/1-reservoir-data/28-archive-2015.html.
16. ESRI. 2018. "ArcGIS Online." accessed March 1,2018.http://www.esri.com/software/arcgis/arcgisonline
17. NASA. 2018. "NASA Surface meteorology and Solar Energy." NASA, accessed March 1,2018. https://eosweb.larc.nasa.gov/cgi-bin/sse/retscreen.cgi?email=rets@nrcan.gc.ca.
18. Short, Walter, Daniel J Packey, and Thomas Holt. 1995. A manual for the economic evaluation of energy efficiency and renewable energy technologies. National Renewable Energy Lab., Golden, CO (United States).
19. OSU. 2018. "River Engineering & Restoration at OSU." Oregon State University learning module on dam planning and operations, accessed March 25,2018. http://rivers.bee.oregonstate.edu/elevation-discharge-power-relationship.
20. Report, Electricity GHG Inventory. 2009-10. GHG Inventory report for Electricity generation and consumption in India.
21. IMD. 2018. Ministry of Earth Sciences,Goverment of India, accessed March 18,2018. http://www.imd.gov.in/Welcome%20To%20IMD/Welcome.php.

11

BLDC Motor Driven Water Pumping System Powered by Solar Photovoltaics (PV)

Dileep Kumar[1]*, Md Tabrez[1], Surya Deo Choudhary[1] and Farhad Muhsin Mahmood[2]

[1]*Department of Electrical and Electronics Engineering, Motihari College of Engineering, Motihari, Bihar, India*
[2]*Department of Physics, College of Science, University of Halabja, Halabja, Iraq*

Abstract
The eradication of fuels and increasing levels of greenhouse gases have fueled research in the area of renewable energy resources. The electricity generated by solar photovoltaics has a very impressive scope in countries such as India. Solar photovoltaics can be used to meet increasing energy demands. In a country like India, which ranks second in farm output in the world, field irrigation largely relies on rain and fossil fuel-operated water pumping systems (WPSs). Solar photovoltaics-powered WPS can be a game changer in the areas of crops, household uses, and industrial applications. This dissertation focuses on the design of a simple, low-cost, and efficient water pumping system, which can offer a good steady state and dynamic performance even with abrupt variations in environmental conditions. This paper leads to the evaluation of popular maximum power point (MPP) techniques so that PV can be operated at the MPP. The MPP operation ensures optimum utilization of the PV array. DC–DC converters were used at the MPP. The BLDC motor is the most efficient drive compared to the DC motors and requires less maintenance owing to the absence of brushes and commutators. The work was validated solely using the MATLAB/Simulink software.

***Keywords*:** BLDC motor, solar PV, MPPT, DC–DC converter

**Corresponding author*: dk18182@gmail.com

Mohammed Aslam Husain, Md Waseem Ahmad, Farhad Ilahi Bakhsh, P. Sanjeevikumar and Hasmat Malik (eds.) Photovoltaic Systems Technology, (211–224) © 2024 Scrivener Publishing LLC

11.1 Introduction

Owing to the tremendous increase in energy usage over the past few decades, the energy crisis has worsened in recent years. Fossil-fuel depletion and environmental problems caused by the consumption of fossil fuels and conventional power generation have emerged as a matter of concern [1]. Significant research on new clean energy sources is ongoing worldwide. These studies have resulted in improved performance and reduced cost [2]. A typical PV module and its exact circuit are shown in Figure 11.1. Solar water pumping has recently gained in popularity. Gas and gasoline-engine-based WPS are not as efficient and generate considerably high amounts of noise and vibrations during operation.

In addition, because fossil fuels are used to operate water pumping systems, they produce a considerably high amount of carbon emissions [3]. However, an electric motor is much cheaper to operate because its electricity costs are very low compared to those of gas and petrol, and it does not produce any carbon emissions [4]. The use of solar power ensures completely pollution-free water pumping. Solar water pumping is an application of a standalone PV system. The reasons for this have been discussed earlier. It is expected that the cost of photovoltaic systems will decrease in the coming years [5]. It is not surprising that the demand for photovoltaics-powered systems will reach a new high level. A BLDC motor-based WPS have not been sufficiently explored. The performance of PV water pumping depends on the performance of the MPP tracking, DC–DC converter, and control of the BLDC motor. A Z-source inverter (ZSI) is used to drive the BLDC motor and MPPT simultaneously [6]. The ZSI-based system operates at high efficiency. However, the control of the ZSI becomes very complex for operating the BLDC and MPP tracking. Methods such as direct torque control and hysteresis current control are used for the precise

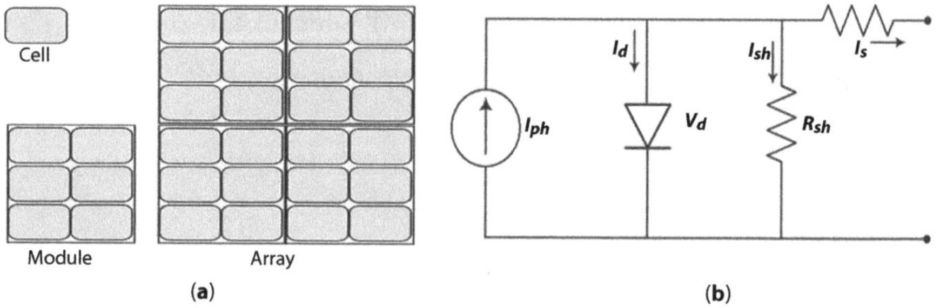

Figure 11.1 (a) Representation of PV cell, module, and array; (b) Equivalent model of PV.

control of the speed performance of BLDC motors [7, 8]. Nevertheless, the precise control of the WPS is not required. The operation of the BLDC motor without the rotor position offers great reliability. The line-voltage zero-crossing technique can be used to achieve the rotor position [9, 10].

In this study, a solar-system-based WPS was designed using a BLDC motor. The overall analysis was based on the MATLAB/Simulink software. The results clearly show that the proposed system is the best choice for different irrigation systems, rather than the conventional WPS.

11.2 Interaction of PV Array and Load

For the output characteristics (i.e., P–V and I–V characteristics) of the PV module, it was pointed out that PV has only one point where power output from PV is maximum [11, 12]. This particular point is known as maximum power point (MPP) and the voltage and current corresponding to the MPP are represented as Vmpp and Impp, respectively. If a load line is drawn at this point, as shown in Figure 11.2, the maximum power is transferred to the load corresponding to the load line. This value of load resistance is denoted by Rmpp.

Figure 11.2(a) shows a PV array connected to variable load RL. The output power of the PV array changes depending on the size and type of the load connected to it. The operating points of the PV array with different load values are shown in Figure 11.2(b). When the load RL is equal to 0Ω (i.e., the PV array terminals are short-circuited), the PV operates at point A. At point A, the current Isc flows from the PV terminals; however, the power output of the PV array is zero because the PV voltage is zero at this

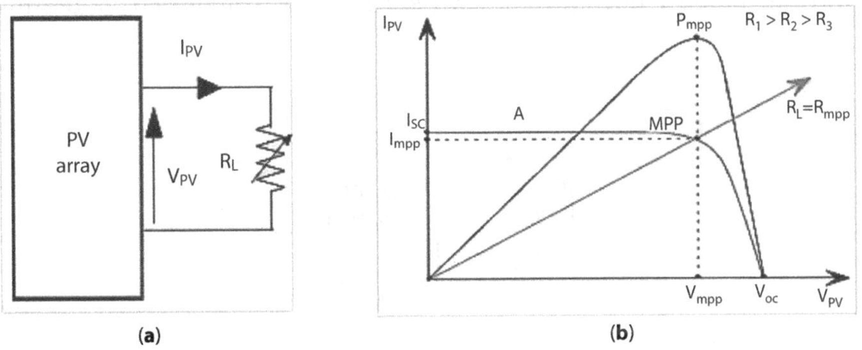

Figure 11.2 (a) PV connected to load; (b) Load characteristic.

point [13, 14]. However, as RL increases to R3, the operating point moves to point B, and the PV power output becomes P_B. The PV power output increased with load resistance until the load resistance became equal to Rmpp. When RL = Rmpp, the PV operates at the MPP (i.e., maximum power point), and the output power attains a maximum possible value Pmpp. As previously discussed, the earlier voltage and current corresponding to the MPP are Vmpp and Impp, respectively. If RL is further increased, the PV power output gradually decreases with an increase in load resistance. The PV power output becomes zero when the load resistances RL becomes infinite. At RL = ∞, the PV array operates at point D, where the PV output voltage is Voc and current is equal to zero. Therefore, the operating point of the PV, and thus the output power of the PV, changes with the load. The load in a power system is highly variable with random variation because generating stations do not have any control over consumers. It is desirable that the load resistance remains equal to the resistance corresponding to the MPP, i.e., Rmpp, so that efficient utilization of PV can be achieved.

11.3 Application of DC–DC Converter for MPPT

DC–DC converters are generally used to step up or down the DC voltage by changing the duty cycle of a high-frequency switch. However, in PV systems, DC–DC converters are used for impedance matching [15, 16]. The most general and commonly used configuration of a standalone PV system is shown in Figure 11.3. The input resistance of the DC–DC converter depends on the duty cycle of the converter. The effective load seen by the PV array can be changed by changing duty cycle of converter. If the load line of the effective load connected to the PV passes through the MPP, then the PV operates at the MPP. This is the main idea behind using

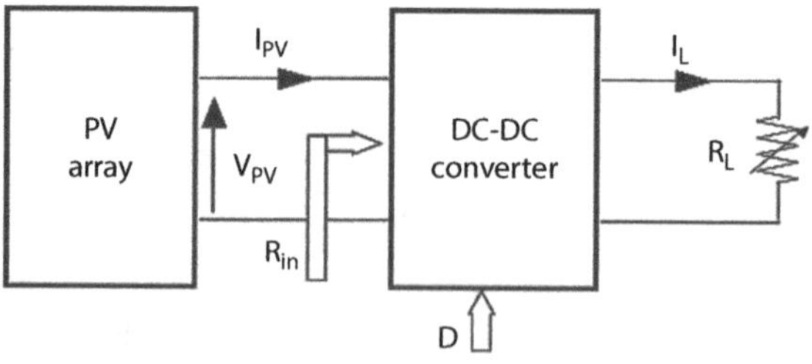

Figure 11.3 PV Array interface with DC–DC converter.

a DC–DC converter between the load and the PV array. Now, we will go through popular DC–DC converter topologies to determine the most suitable converter for PV systems.

11.4 Three-Phase BLDC Motor

A block diagram of the BLDC motor drive is shown in Figure 11.4. In a BLDC motor, Hall sensors are placed on the stator to sense and provide rotor position feedback. The three Hall sensors provided the rotor position in the form of logic (0 or 1) signals. The Hall sensor output was decoded by a Hall decoder circuit to achieve synchronized commutation.

The decoder circuit is designed to operate the inverter in 120° mode of operation. In the 120° mode of operation, each switch of the inverter operates electrically at 120° in every cycle. This logical table was designed based on the rotor position obtained from the Hall signals. Only two windings of the BLDC motor were energized simultaneously. Unlike a PMSM, all three windings are excited simultaneously.

The switches of the power converter are divided into two groups: positive and negative. The upper switches in each leg, such as S_1, S_3, and S_5, belong to the positive group and the lower switches (S_2, S_4, and S_6) belong to the negative group. At a time, two switches from two different phases, one from the positive group and one from the negative group, are put in the condition mode. Based on the above switching table, the operation of the inverter over one cycle can be divided into six segments which are as following:

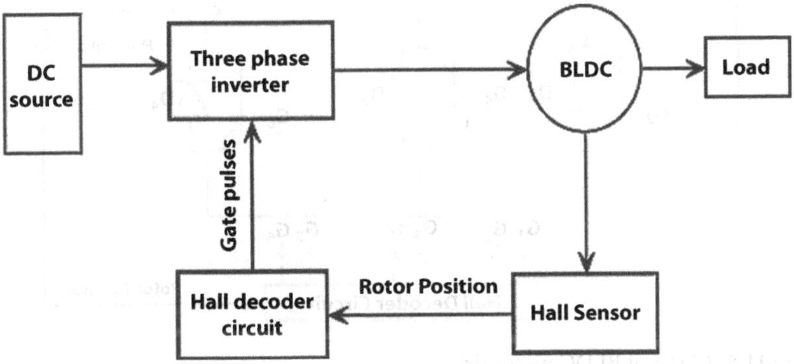

Figure 11.4 Layout of the BLDC motor drive.

Mode I (0°–60°): During this period switch, S_1 and S_4 are ON. In this segment, two phases, namely phase 'a' and 'b' are active, i.e., source supplies power to motor through phase 'a; and 'b.' Phase 'c' remains open circuited, i.e., current in phase 'c' is zero whereas the positive current flows through the phase 'a' and the current through phase 'b' is negative as shown in Figure 11.5. Switch S_4 was turned off at the end of this segment, i.e., at 60° electrical.

Mode II (60°–120°): During this period switch, S_1 remains in ON state and Switch S_6, negative group switch of the third leg, is turned on after commutation of negative group switch S_4 of the second leg. This sequence is referred to as the first commutation sequence. Phase 'b' current becomes zero, and phase 'c' current comes in picture. Phase 'a' current remain same as in the previous segment, i.e., 0°–60° electrical.

Mode III (120°–180°): Phase 'a' current becomes zero due to commutation of switch S_1 at 120°. Switch S3 is turned ON during the 120°–180° and phase 'b' current is positive. The current in Phase 'c' remained the same.

Mode IV (180°–240°): During this mode phase 'c' current become zero due to commutation of switch S6. Switch S3 continues in the ON state. Switch S2, negative group switch is turned ON and phase 'a' current provides return path for phase 'b' current.

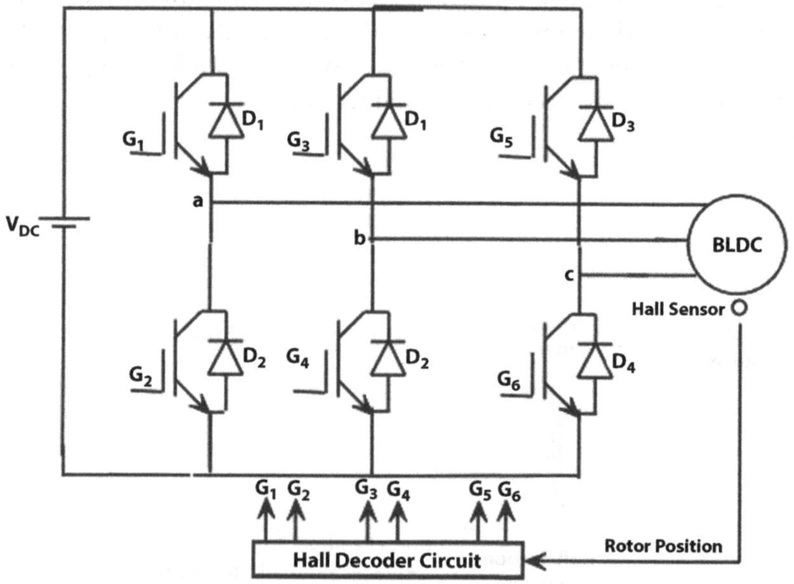

Figure 11.5 Classical BLDC motor drive.

Mode V (240°–300°): **V:** During this period switch S3 is off and phase 'b' current is zero. Switch S5 is turned ON. Phase 'a' current remains unchanged and provides return path for new phase 'c' current.
Mode VI (300°–360°): In this mode phase 'c' current continue to flow through S5. However, the current in phase 'a' becomes zero, because switch S2 is commutated. Phase 'c' current returns passes through phase 'b' (switch S4).

From the above discussion, it can be concluded that the inverter operates in 120° mode. Only two switches, one from the upper group and the other from the lower group, but with different legs, are conducting. After 60°, one switch out of the two turns off, and the other switch operates for the next 60°. For example, suppose that switches S1 and S4 are conducting at 60°. After 60°, switch S4 is turned off, switch S6 is turned on, and switch S1 is continues to be on state for the next 60°. When switches S1 and S4 are ON, the current flowing in phase 'a' is positive and current in phase 'b' is negative. After 60° phase 'b' transfers current to phase 'c' and current in phases 'b' becomes zero and current across phase 'c' is negative. During this period, the phase 'a' current remains positive. Next, all the commutation sequences are similar to those above. PWM techniques are sometimes used to control the voltage and current across a BLDC motor terminal. In this method, the turn-on time of the switch depends on the duty cycle.

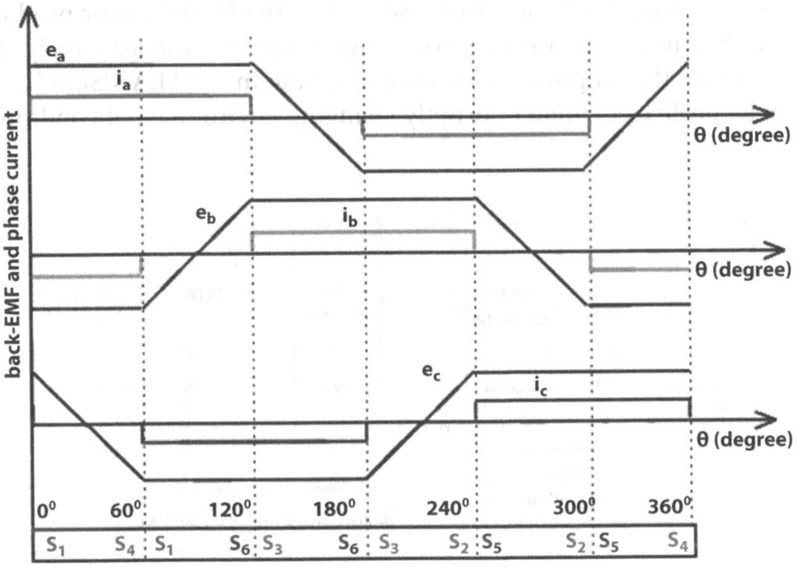

Figure 11.6 Wave shape back-EMF and current.

Two PWM techniques, feedback mode and freewheeling mode, are used to control the voltage and current. The trapezoidal back-EMF and rectangular current are shown in Figure 11.6.

11.5 Simulation of Suggested Technique

BLDC motor-assisted WPS have gained increasing attention. The BLDC motor is more efficient than the DC motor because of the absence of a brush and commutator assembly and the of use of a permanent magnet. In addition, it has an almost negligible maintenance requirement. The initial cost of the BLDC motor is a major issue. However, in the longer run, the BLDC motor compensates for the initial cost using the cost of energy saved by its use. In Figure 11.7, in the WPS, we do not need precise control of the motor for obvious reasons. WPS supplied by photovoltaics are gaining importance in the present scenario. This is because of the danger of eradicating fossil fuels in the coming future and because of the alarming level of pollution and global warming caused by the consumption of fossil fuels.

The PV-fed WPS may or may not have electricity storage arrangements. Water pumping with storage has a high initial cost but can operate even when the sun is not there. However, our focus is on PV water pumping without electric storage. The voltage output of the PV array changes with solar irradiance and temperature. Therefore, the DC-link voltage, which works as an input to the inverter, also varies. The BLDC motor produces a good dynamic response even for a highly variable voltage supply. The simulation of the proposed WPS was carried out in MATLAB/Simulink to verify its proficiency under abruptly changing environmental conditions,

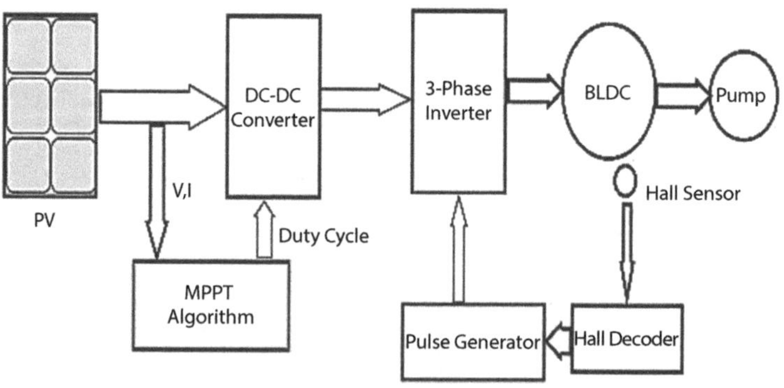

Figure 11.7 PV-fed BLDC-driven water-pumping system.

as shown in Figure 11.8. The specifications of the BLDC motor used for the simulations are given in the Appendix.

The simulation results for PV irradiance and temperature are shown in Figure 11.9. The simulation results are presented in Figure 11.10. The PV array MPP was tracked accurately by the INC MPPT by perturbation in the duty cycle. A small step size has produced satisfactory results in both MPP tracking and BLDC motor start. The BLDC motor variables are shown in Figure 11.10(b). The motor takes a very small current at the start, which is even lower than the full-load current. Figure 11.10(c) shows the speed and torque of the BLDC motor and the pump. The speed in the steady state had

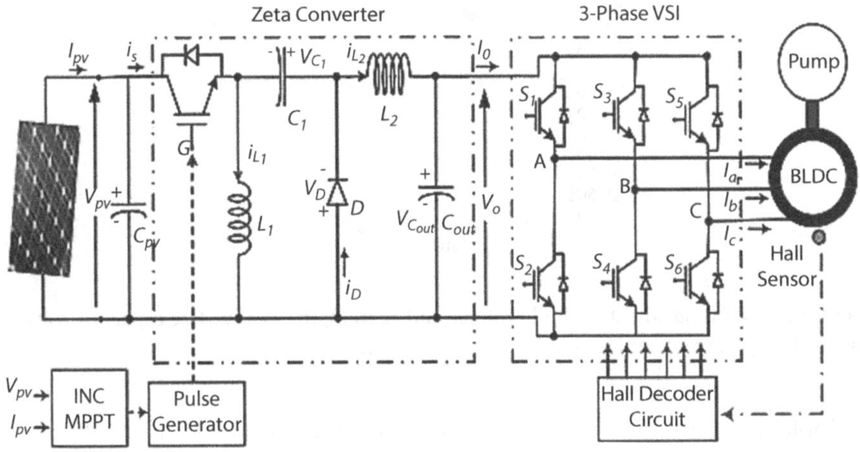

Figure 11.8 PV-fed BLDC-assisted WPS.

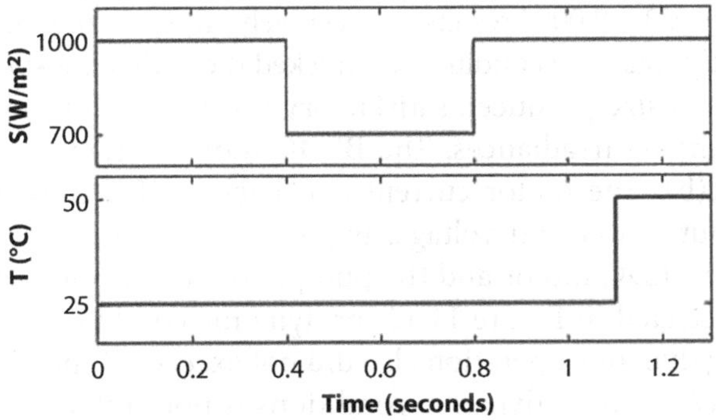

Figure 11.9 Irradiance and temperature variation.

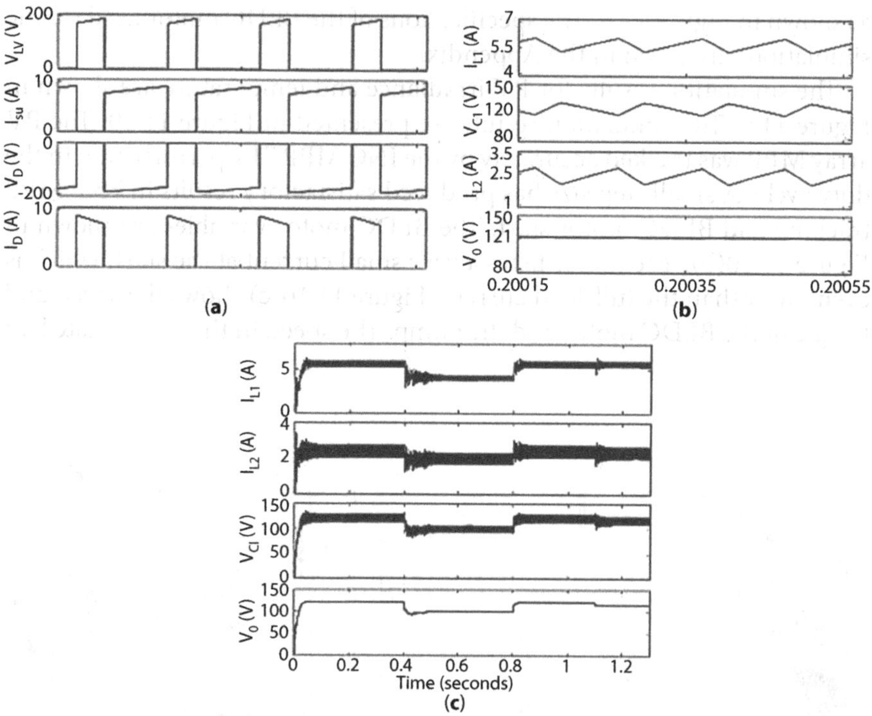

Figure 11.10 (a) Steady-state diode voltage and current, (b) steady-state capacitor voltage and inductor current, and (c) transient performance.

a ripple that remained below 30 rpm. This ripple in speed for a WPS is not problematic.

The dynamic performance results of the simulation are shown in Figure 11.11. The PV array MPP is tracked accurately by INC MPPT by perturbation in the duty cycle, even with a very abrupt change in irradiance. A small step size ΔD = 0.0005 results in a relatively large tracking time. However, under abruptly changing conditions, it tracked the MPP almost instantly.

A small step size produced satisfactory results in MPP tracking under abruptly changing irradiances. The BLDC motor variables are shown in Figure 11.11(b). The motor current and back-EMF follow variations in the PV output power and voltage. Figure 11.11(c) shows the speed and torque of the BLDC motor and the pump. The results for various parameters are illustrated in Figure 11.12 for dynamic conditions. The ripple in speed during dynamic operation also did not exceed 30 rpm. This ripple in speed for a WPS under dynamic conditions is not problematic. Thus, the dynamic performance of the proposed system is used.

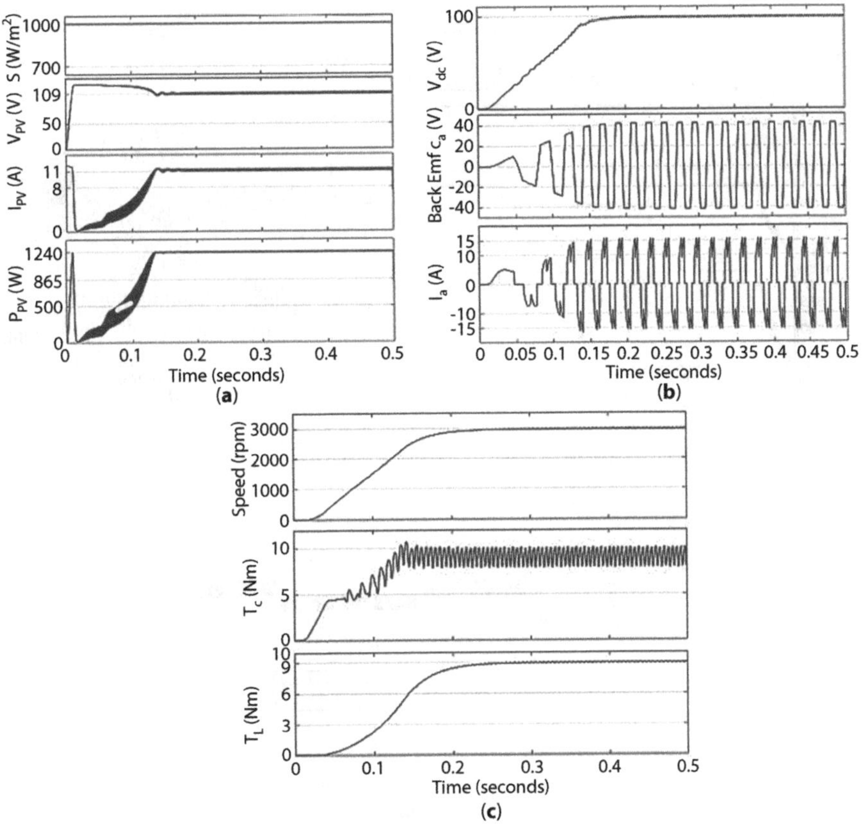

Figure 11.11 (a) PV array, (b) BLDC motor variables, and (c) speed and torque of WPS.

11.6 Conclusion

Water pumping is required for irrigation, household use, and industrial applications. Eradication of fossil fuels and dangerously increasing pollution have fueled research in the area of renewable energy-powered WPS. Electricity generated by solar photovoltaics is a potential option for supplying power to water-pumping systems. In this study, solar-powered WPS was suggested. The proposed system uses a BLDC motor to drive a centrifugal pump. The proposed system was simulated in MATLAB/Simulink to validate the performance of varying the irradiance and temperature. By adopting a small step size ($\Delta D = 0.0005$), soft starting of the BLDC motor was made possible for MPP tracking. Fundamental frequency switching

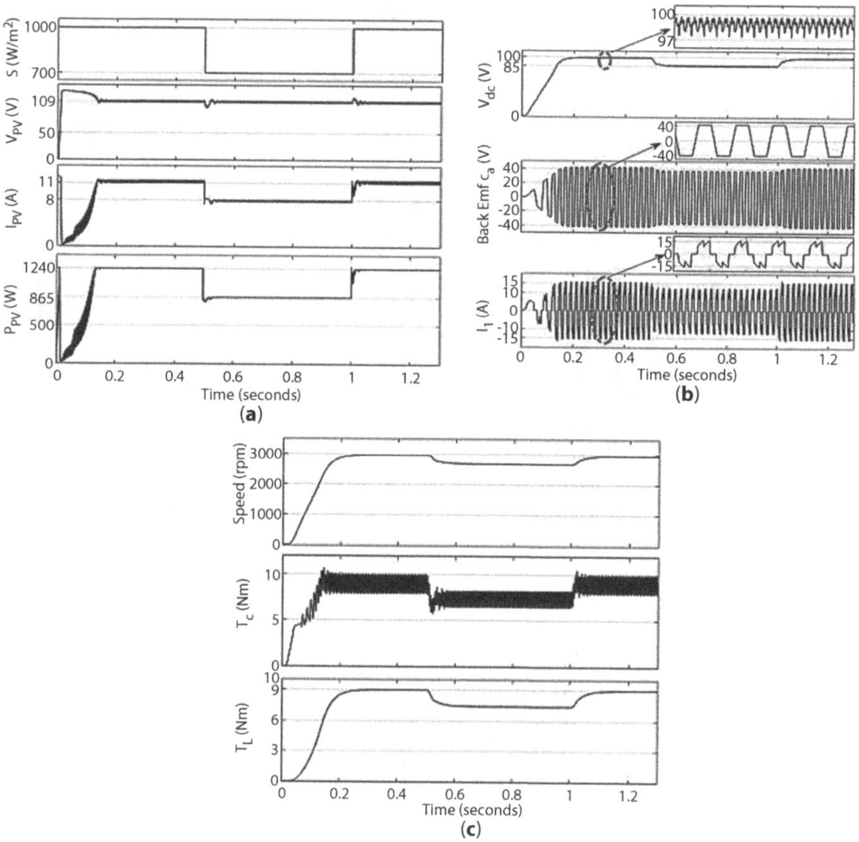

Figure 11.12 Dynamic performance of (a) PV array, (b) BLDC motor variables, and (c) speed and torque of WPS.

for 3 phase VSI was used to minimize the switching losses in the inverter. Simulations results were obtained.

References

1. Sen and B. Singh, "Peak current detection starting based position sensorless control of BLDC motor drive for PV array fed irrigation pump," in *Proc. IEEE Int. Conf. Environ. Elect. Eng. Ind. Commercial Power Syst. Europe (EEEIC/I&CPS Europe)*, 2019, pp. 1–6.
2. S. Jain, A. K. Thopukara, R. Karampuri, and V. T. Somasekhar, "A single-stage photovoltaic system for a dual-inverter-fed open-end winding induction

motor drive for pumping applications," *IEEE Trans. Power Electron.*, vol. 30, no. 9, pp. 4809–4818, Sep. 2015.
3. L. An and D. D. Lu, "Design of a single-switch DC/DC converter for a PV-battery-powered pump system with PFM+PWM control," *IEEE Trans. Ind. Electron.*, vol. 62, no. 2, pp. 910–921, Feb. 2015.
4. J. V. M. Caracas, G. d. C. Farias, L. F. M. Teixeira, and L. A. d. S. Ribeiro, "Implementation of a high-efficiency, high-lifetime, and low-cost converter for an autonomous photovoltaic water pumping system," *IEEE Trans. Ind. Appl.*, vol. 50, no. 1, pp. 631–641, Jan./Feb. 2014.
5. T.-H. Kim and M. Ehsani, "Sensorless control of the BLDC motors from near- zero to high speeds," *IEEE Trans. Power Electron.*, vol. 19, no. 6, pp. 1635–1645, Nov. 2004.
6. S. Dusmez, A. Khaligh, M. Krishnamurthy, E. Ugur, and M. Uzunoglu, "Sensorless control of BLDCs for all speed ranges with minimal components," in *Proc. Inter. Aegean Conf. Elect. Machines Power Electron. Electromotion, Joint Conf.*, 2011, pp. 626–631.
7. J. Park, S. Hwang, and J. Kim, "Sensorless control of brushless DC motors with torque constant estimation for home appliances," *IEEE Trans. Ind. Appl.*, vol. 48, no. 2, pp. 677–684, Mar./Apr. 2012.
8. J. Feng, K. Liu, and Q.Wang, "Scheme based on buck-converter with three-phase H-bridge combinations for high-speed BLDC motors in aerospace applications," *IET Electr. Power Appl.*, vol. 12, no. 3, pp. 405–414, Mar. 2018.
9. A. Ahfock and D. Gambetta, "Sensorless commutation of printed circuit brushless direct current motors," *IET Electr. Power Appl.*, vol. 4, no. 6, pp. 397–406, Jul. 2010.
10. R. Kumar and B. Singh, "Buck-boost converter fed BLDC motor drive for solar PV array based water pumping," in *Proc. IEEE Inter. Conf. Power Electron., Drives Energy Syst. (PEDES)*, 2014, pp. 1–6.
11. G. N. A. Maranhão, A. U. Brito, J. T. Pinho, J. K. S. Fonseca, A. M. Leal, and W. N. Macêdo, "Experimental results of a fuzzy controlled variablespeed drive for photovoltaic pumping systems: A review," *IEEE Sens. J.*, vol. 16, no. 9, pp. 2854–2864, May 2016.
12. B. Singh and R. Kumar, "Simple brushless DC motor drive for solar photovoltaic array fed water pumping system," *IET Power Electron.*, vol. 9, no. 7, pp. 1487–1495, 2016.
13. S. Sheik Mohammed, D. Devaraj, and T. P. Imthias Ahamed, "Maximum power point tracking system for stand alone solar PV power system using adaptive neuro-fuzzy inference system," in *Proc. Biennial Inter. Conf. Power Energy Syst.: Towards Sustain. Energy (PESTSE)*, 2016, pp. 1–4.
14. A. Gupta, P. Kumar, R. K. Pachauri, and Y. K. Chauhan, "Performance analysis of neural network and fuzzy logic based MPPT techniques for solar PV systems," in *Proc. 6th IEEE Power India Inter. Conf. (PIICON)*, Delhi, India, 2014, pp. 1–6.

15. K. Lee, D. Kim, B. Kim, and B. Kwon, "A novel starting method of the surface permanent-magnet BLDC motors without position sensor for reciprocating compressor," *IEEE Trans. Ind. Appl.*, vol. 44, no. 1, pp. 85–92, Jan./Feb. 2008.
16. R. Kumar and B. Singh, "Single stage solar PV fed brushless DC motor driven water pump," *IEEE J. Emerg. Sel. Top. Power Electron.*, vol. 5, no. 3, pp. 1377–1385, Sep. 2017

Appendix

Peak Power P_{mpp}—310.15 W, Open Circuit Voltage (Voc)—64.4 V, Voltage at MPP, V_{mpp}—54.7 V, Short Circuit Current, I_{sc}—6.05 A, Current at MPP, I_{mpp}—5.67 A, Number of Cells Per Module-96, Number of Series Connected Modules-2, Number of Parallel Connected Modules-2.

12

Hybrid Photovoltaic/PEM Fuel Cell Driven Water Pumping System for Agricultural Application: Overview, Challenges and Future Perspectives

Rupendra Kumar Pachauri[1], Renu Mavi[2], Ahmad Faiz Minai[1,3*], Shubham Tiwari[4] and Shashikant[5]

[1]*Electrical Cluster, School of Advanced Engineering, UPES, Dehradun, Uttarakhand, India*
[2]*Department of Chemistry, Swami Vivekanand Subharti University, Meerut, Uttar Pradesh, India*
[3]*Electrical Engineering Department, Integral University, Lucknow, Uttar Pradesh, India*
[4]*Department of Sustainable Energy Transition, School of Environment, Resources and Development, Asian Institute of Technology, Bangkok, Thailand*
[5]*Electrical Engineering Department, Babu Banarasi Das University, Lucknow, Uttar Pradesh, India*

Abstract
Proton exchange membrane fuel cell (PEMFC) and solar photovoltaic (PV) systems have potential applications in agriculture, particularly in powering irrigation systems, greenhouses, and other farm equipment. They can also be used to generate electricity for rural communities and power electric tractors and other vehicles used in farming. One major challenge in using these technologies in agriculture is the high installation and maintenance costs. Additionally, the performance of these systems can be affected by weather conditions such as cloud cover and extreme temperatures. Despite these challenges, there is growing interest in using PEMFC and solar PV systems in agriculture to reduce dependence on fossil fuels and increase energy efficiency. In the future, the cost of these systems is expected to decrease, making them more affordable for farmers. In addition, advances in materials science and energy storage technology may improve the performance

**Corresponding author*: fzminai@gmail.com

Mohammed Aslam Husain, Md Waseem Ahmad, Farhad Ilahi Bakhsh, P. Sanjeevikumar and Hasmat Malik (eds.) Photovoltaic Systems Technology, (225–250) © 2024 Scrivener Publishing LLC

and durability of these systems, making them more viable for agricultural use. In summary, FPEMCs and solar PV systems have potential applications in agriculture; however, the cost- and weather-dependent nature of these systems remains a major challenge. However, the future prospects of these systems look bright as the cost is expected to decrease and technology is expected to improve. In the present study, a mathematical study and MATLAB/Simulink modeling of a hybrid PV/PEMFC power-driven water pumping system were investigated for transient behavior. In addition, hydrogen storage for PEMFC operation is explored along with the challenges, pros, and cons of the present study. The induction motor was driven at 1,728 rpm through PV/FC power assistance.

Keywords: Fuel cell, solar photovoltaic system, renewable energy, agricultural system, water transport

12.1 Introduction

Farming ensures that people in a nation have access to enough food to keep themselves healthy and productive. In addition to its obvious role in producing food and revenue, agriculture also plays a significant supporting role in the economy by facilitating the development of related sectors, such as processing and transportation [1]. Potentially, large gains in terms of income, poverty, and access to essential amenities such as schools and hospitals may be made in rural areas by focusing on agriculture. Sustainable agriculture practices can help conserve natural resources, reduce greenhouse gas emissions, improve the health of the environment, and mitigate the effects of climate change. Agricultural exports can provide a country with a valuable source of foreign income and can help improve its trade balance. Agriculture plays a critical role in the development of a nation, providing food and economic opportunities and a foundation for sustainable and equitable growth [2].

Renewable energy-powered irrigation systems use energy from sources such as solar, wind, or hydropower to operate pumps and control systems for irrigation. This can improve efficiency, and reduce the carbon footprint of agriculture, and provide a more reliable and sustainable source of energy for irrigation in areas with limited access to grid power [3]. Examples of renewable-energy-powered irrigation systems include solar-powered pumps for pumping water from wells, wind turbines for pumping water from rivers, and micro-hydro systems for generating electricity for pumps. These systems can be particularly beneficial for small-scale farmers and rural communities where access to grid electricity is limited [4]. The use of renewable energy sources in irrigation systems is important for several reasons:

- **Energy independence:** By relying on renewable energy sources, farmers can reduce their dependence on fossil fuels and energy costs.
- **Sustainability:** Renewable energy sources are clean and sustainable, reducing the carbon footprint of the agricultural sector and contributing to the fight against climate change.
- **Rural electrification:** Access to electricity is limited in many rural areas and the use of renewable energy sources can help provide power to these communities, thereby improving their quality of life and economic opportunities.
- **Water conservation:** Many traditional irrigation systems are powered by diesel or electricity, which are often expensive and wasteful. Renewable energy sources can help to reduce water consumptionand preserve valuable resources.
- **Improved crop yields:** By using renewable energy sources in power irrigation systems, farmers can provide their crops with a consistent and reliable water supply, leading to improved yields and increased profits.

The use of renewable energy in irrigation systems is crucial for the long-term health of the agricultural industry in any nation [5].

In this context, FC and PV systems can play important roles in agricultural systems, providing energy for various applications, including:

- **Irrigation systems:** powering pumps to extract and distribute water
- **Greenhouses:** supplying energy for lighting and temperature control
- **Livestock operations:** providing power for heating and cooling systems
- **Agricultural machinery:** powering tractors, combines, and other equipment
- **Rural electrification:** supplying energy to remote farm communities.

FC and PV systems-based hybrid power generation offers the advantage of being a highly efficient and reliable power backup at low emissions and environmentally friendly solutions [6]. Figure 12.1 shows a schematic

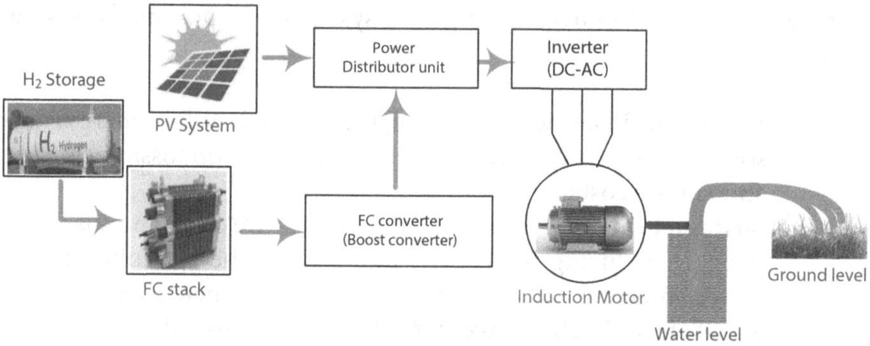

Figure 12.1 Schematic diagram of PV/FC powered water pumping system.

diagram of a hybrid FC/PV power generation system that represents a typical water-pumping system. Table 12.1 explores different types of FC systems, along with the required reactant fuel supply for efficient operation and the electrical operating parameters that need to be met.

The present chapter is comprised of the following sections to describe the study of the FC/PV-powered water pumping system: *Section 12.1* explores the introduction of the chapter theme, and the subsequent section is followed by the literature review. Furthermore, modeling of FC/PV power generation systems and irrigation system applications are explored

Table 12.1 FC types, fuel required and applications [7, 8].

FC types	PEMFC	AFC	PAFC	MCFC	SOFC
Fuel required	H_2	H_2	H_2	H_2, CO, CH_4	CH_4
Efficiency (%)	53–58	60	>40	45–47	35–43
Op. Temp (°C)	50–100	90–100	150–200	600–700	800–1,000
Switching time	Second	Second	Hours	>hours	>hours
Applications	Hospital, EVs Agriculture	• Military • Space	• Residential power backup	• Marine • Railway	• Residential power backup

in *Sections 12.2* and *12.3*. Moreover, the methodologies and limitations for hydrogen fuel generation and storage methods are elaborated in *Section 12.4*. In addition, *Sections 14.5* and *14.6* are more about the future scope of renewable energy irrigation system applications and their pros and cons. Moreover, *Section 12.7* explores the performance investigation of a PV/FC-powered water pumping system through the MATLAB/Simulink study. At the end of the chapter, *Section 12.8* concludes the chapter.

12.2 Mathematical Modeling

12.2.1 PEMFC System

A PEMFC was modeled by simulating its internal physical, chemical, and electrochemical operations. These models may be used to forecast how a PEMFC functions in each scenario and to optimize the fuel cell design for a certain application [9, 10]. Furthermore, these models are crucial for the research and development of PEMFCs technologies. Figure 12.2 shows a schematic representation of the FC.

Eq. (12.1) provides a theoretical prediction of the maximum voltage of FC (V_{fc}) that can be produced the FC, assuming ideal conditions [11]. In practice, the actual voltage of a PEMFC is lower than the Nernst voltage because of various losses, such as activation, concentration, and ohmic voltage. In the case of a PEMFC, the Nernst voltage can be modeled using the Nernst equation as shown in Eq. (12.2),

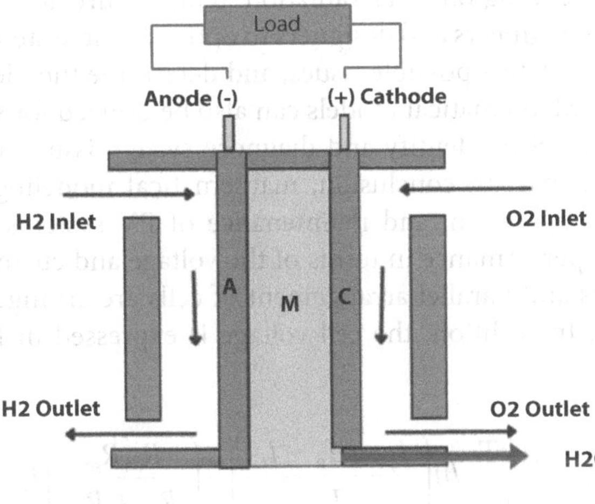

Figure 12.2 FC: Schematic diagram [10].

$$V_{fc} = E - V_{act} - V_{conc} - V_{ohm} \tag{12.1}$$

$$E = E° - (RT / nF)\ ln\left(\frac{P_{O2}}{P_{H2}}\right) \tag{12.2}$$

where E = the Nernst voltage; E° = the standard electrode potential; R = gas constant; T = temperature (K); n = electron number; F = Faraday's constant; and the reactant (hydrogen and oxygen) fuel pressures are P_{H2} and P_{O2}.

The chemical reaction that must take place to accomplish FC operation with the proper reactant flow pressures at the anode and cathode is represented by Eq. (12.3) [12], which may be found below,

$$\left.\begin{array}{l} Anode: H_2 = 2H^+ + 2e^- \\ Cathode: \frac{1}{2}O_2 + 2H^+ + 2e^- = H_2O \\ Cell: H_2 + \frac{1}{2}O_2 = H_2O \end{array}\right\} \tag{12.3}$$

12.2.2 PV System

The performance of the PV system may be predicted and optimized using mathematical modeling. These models can simulate the electrical output of a system depending on solar radiation, temperature, and shading [13, 14]. This allows engineers and designers to optimize the system setup, forecast its behavior, detect possible issues, and determine the ideal operating circumstances. Mathematical models can also be utilized for system monitoring and control to identify and diagnose system issues and optimize system performance. In conclusion, mathematical modeling is essential for the design, operation, and maintenance of PV systems [15–21]. To achieve higher performance in terms of the voltage and current of the PV system, a series and parallel arrangement of cells are arranged, as shown in Figure 12.3. In addition, the cell voltage is expressed in Eq. (12.4) as follows,

$$V_c = \frac{AkT_C}{e} ln\left(\frac{I_{ph} + I_d - I_C}{I_d}\right) - \left(\frac{R_{sh}R_{se}}{R_{sh} + R_{se}}\right) I_C \tag{12.4}$$

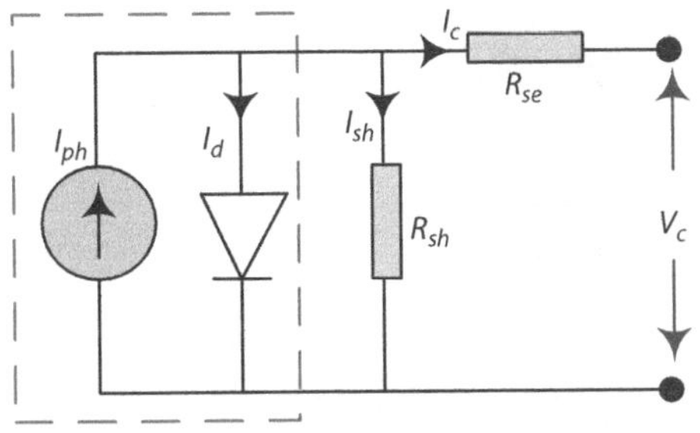

Figure 12.3 PV module configuration in PV array (m × n).

where V_c = cell voltage; A, k, T_C, e are represented for ideality factor, Boltzman constant, cell temperature and electron charge; I_{ph}, I_d, I_C stands for photo current, diode saturation current and cell current; R_{sh} and R_{se} represented for shunt and series resistances.

12.3 MATLAB/Simulink Study of Hybrid FC/PV Powered Water Pumping System

The hybrid FC/PV-driven water pumping system shown in Figure 12.4(a) is the result of MATLAB/Simulink analysis. The hybrid FC/PV power generation, power electronics system, and electric motor can all be simulated in MATLAB/Simulink, allowing users to analyze their dynamic behavior under changing loads. As shown in Table 12.2, even a novice may be inspired to create the basic mathematical modeling required to create a hybrid PV/FC-driven water pumping system. In the Figures 12.4(b) and 12.4 (c), the sub-systems of the hybrid PV/FC system are given as,

Studying the stator current, settling response of rotor speed, and electromagnet (EM) torque of the aided electric drive system, as well as the PV/FC power production and behavior of the hybrid PV/FC-powered electric drive system integrated water pumping system, yields the following insights. The transient responses of the hybrid PV/FC and inverter voltages are shown in Figures 12.5 and 12.6 as follows,

The transient response of the electric drive system in terms of the stator current, rotor speed (1,780 rpm), and electromagnetic torque is shown in Figure 12.7 as follows,

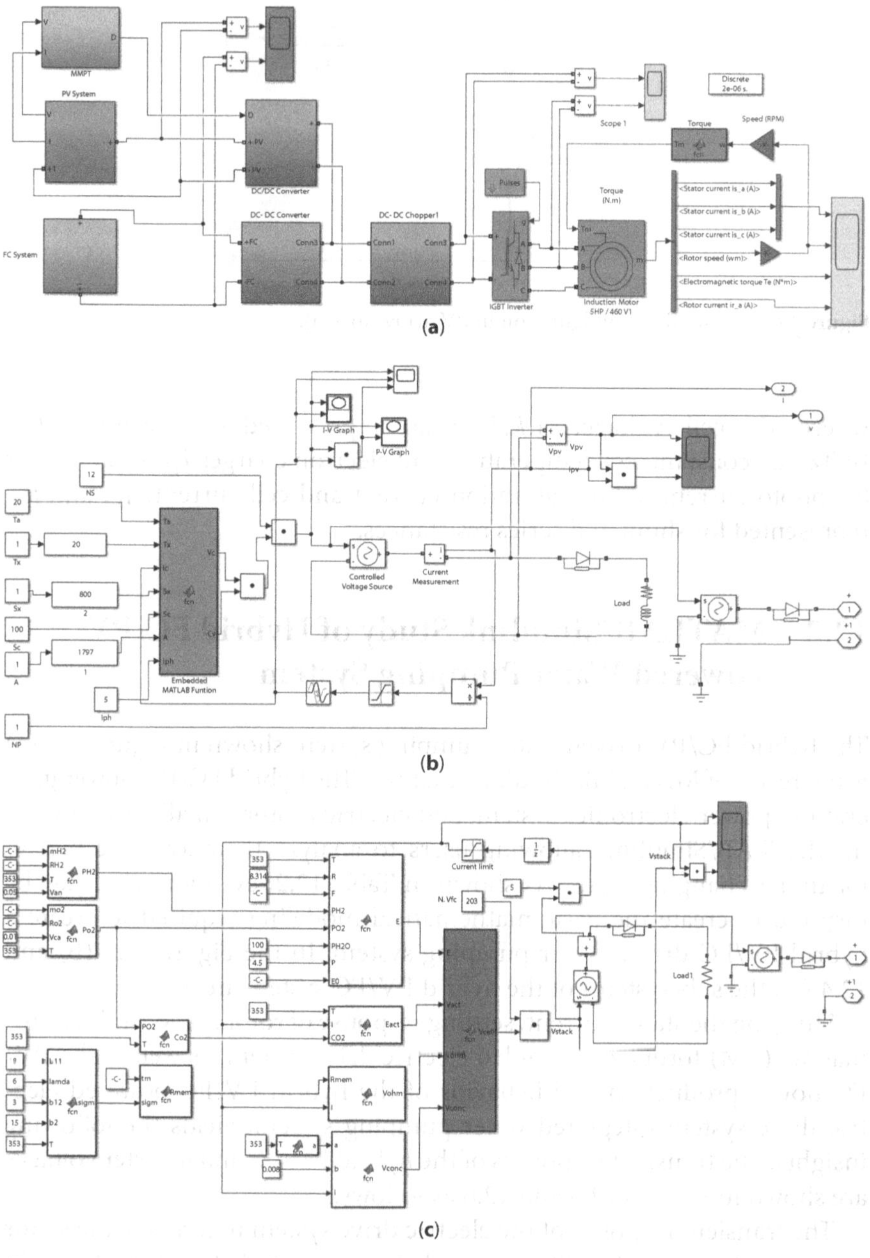

Figure 12.4 (a) MATLAB/Simulink modeling of hybrid PV/FC driven water pumping system, (b) PV system, and (c) PEMFC system.

Table 12.2 Mathematical modeling of PEMFC and PV system [7, 8].

Performance parameters	Mathematical modeling	Abbreviation
Fuel cell voltage (V_{fc}) and Nernst voltage	$V_{fc} = E - V_{act} - V_{conc} - V_{ohm}$, $E = E_0 + \frac{RT}{2F}\left(\ln\left\{\left(\frac{P_{H_2}}{P_{H_2O}}\right)(P_{O_2})^{\frac{1}{2}}\right\}\right)$	T, cell temperature; F, Faraday constant, $P_{H2}, P_{O2}, and\, P_{H2O}$, Hydrogen, oxygen and water flow pressure; R, gas constant; E_0, reference voltage; E, open circuit potential; V_{act}, activation voltage; V_{Conc}, concentration voltage, V_{ohm}, ohmic voltage;
Activation voltage (V_{act}), and oxygen Concentration	$Conc.O_2 = \left(\frac{Po_2}{5.08\times10^6 \times e^{\left(-498/T\right)}}\right)$, $V_{act} = -k_1 + k_2T - \left(\ln\left\{\frac{(I_{fc})^{k_3}}{(Conc.O_2)^{k_4T}}\right\}\right)$	$K_1, K_2, \& K_3$, constant; I_{fc}, fuel cell current; Conc. O_2, oxygen concentration.
Ohmic, concentration voltages, and membrane resistance	$V_{ohm} = I_{fc}R_{mem}$, $V_{conc} = F_1 - F_2(T-273)e^{(F_3 I)}$, $R_{mem} = \frac{t_m}{\sigma}$	t_m, membrane thickness; σ, fuel conductivity; R_{mem}, Membrane resistance; $F_1, F_2, \& F_3$, Coefficients

(*Continued*)

Table 12.2 Mathematical modeling of PEMFC and PV system [7, 8]. (*Continued*)

Performance parameters	Mathematical modeling	Abbreviation
Hydrogen and oxygen fuel pressure	$P_{H_2} = \left(\frac{(m_{H_2} R_{H_2})}{V_{anode}}\right) T, P_{O_2} = \left(\frac{(m_{O_2} R_{O_2})}{V_{cathode}}\right) T$	V_{anode}, anode voltage; m_{H2}, hydrogen molar flow rate; R_{H2}, hydrogen gas constant; $V_{Cathode}$, cathode voltage; m_{O2}, oxygen molar flow rate; R_{O2}, oxygen gas constant
PV cell voltage	$V_c = \frac{AkT_C}{e} ln\left(\frac{I_{ph} + I_d - I_C}{I_d}\right) - \left(\frac{R_{sh} R_{se}}{R_{sh} + R_{se}}\right) I_C$	V_c, cell voltage; A, ideality factor; k, Boltzmann constant; T_c, cell temperature; I_{ph}, photocurrent; I_d, saturation current; I_c, cell current; R_{sh}, shunt resistance; R_{se}, series resistance.

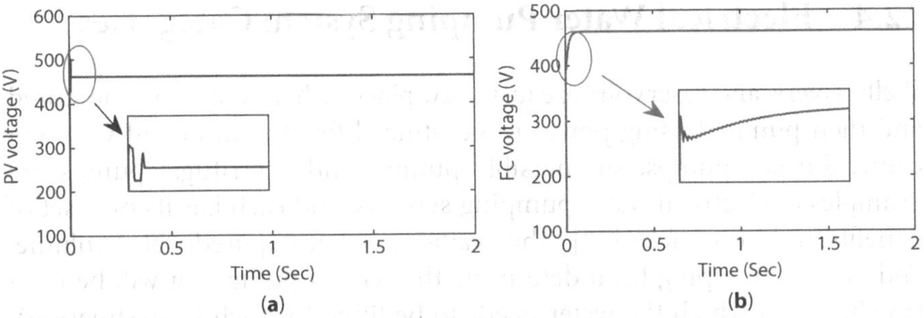

Figure 12.5 Transient response: (a) PV voltage and (b) FC voltage.

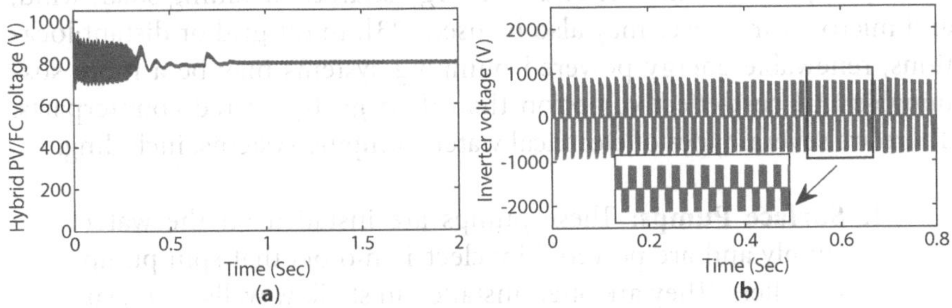

Figure 12.6 Voltage transient response: (a) Hybrid PV/FC voltage and (b) Inverter voltage.

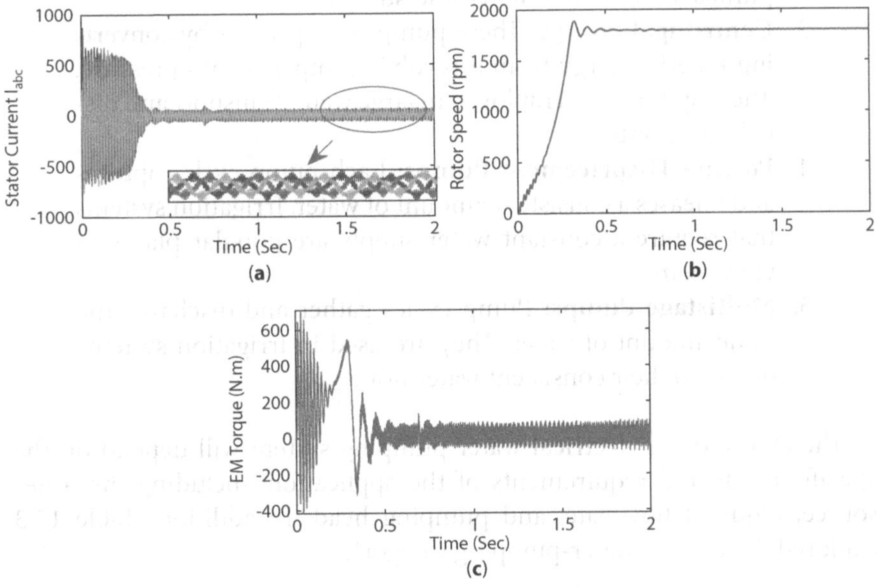

Figure 12.7 Transient response: (a) Stator current, (b) Rotor speed, and (c) EM torque.

12.4 Electrical Water Pumping System Categories

Wells, rivers, and reservoirs are just a few places where water may be stored and then pumped using power to be utilized for irrigation and watering cattle. Surface pumps, submersible pumps, and centrifugal pumps are examples of electrical water pumping systems, and each has its own set of benefits and drawbacks [22]. The available water, required water volume, and desired pumping head determine the type of pump that will be used (the height at which the water needs to be lifted to reach its destination). While grid electricity is the most common source of power for electrical water pumping systems, renewable energy sources including solar, wind, and micro-hydropower may also be used [23]. In off-grid or distant locations, renewable energy-powered pumping systems may be a more sustainable and cost-effective option than their grid-powered counterparts. There are several types of electrical water-pumping systems, including:

1. **Surface Pumps:** These pumps are installed on the water supply and are powered by electric motors that spin pump impellers. They are often installed in shallow wells and near the surface of water sources [24].
2. **Submersible Pumps:** Deeper wells and reservoirs employ totally submerged pumps. They perform better than surface pumps in boreholes, wells, and sumps.
3. **Centrifugal Pumps:** These pumps pump water by converting kinetic energy from a revolving impeller into pressure. These systems are employed in large water transport and distribution systems.
4. **Positive Displacement Pumps:** Each pump cycle captures and releases a consistent amount of water. Irrigation systems that require a constant water supply are popular places to view them.
5. **Multistage Pumps:** Pump cycles gather and discharge the same amount of water. They are used in irrigation systems owing to their consistent water flow.

The choice of an electrical water pumping system will depend on the specific needs and requirements of the application, including the water source, required flow rate, and pumping head. In addition, Table 12.3 explored the electric water-pumping categories.

Table 12.3 Electric water pumping categories [25, 26].

Type	Major components	Role of components
Surface pumps	• Electric Motor • Pump Impeller • Pump Housing • Inlet and Outlet Valves • Suction and Discharge Piping • Drive Shaft • Control Panel • Protection Devices • Foundation	Surface pumping systems have the following main components: 1. **Electric Motor**: Provides the power to drive the pump 2. **Pump Impeller**: Moves the water by creating a flow and generating pressure. 3. **Pump Housing**: Surrounds the pump impeller and contains the working components of the pump. 4. **Inlet and Outlet Valves**: Control the flow of water into and out of the pump. 5. **Suction and Discharge Piping**: These connect the pump to the water source and the destination, respectively. 6. **Drive Shaft**: This connects the electric motor to the pump impeller. 7. **Control Panel**: This houses the electrical components and control systems for the pump, including switches, fuses, and wiring. 8. **Protection Devices**: These include overload protection, dry-run protection, and pressure switches to ensure the safe and efficient operation of the pump. 9. **Foundation:** A solid foundation is necessary to support the pump and motor and to prevent damage due to vibration. Check valves, pressure tanks, and filters may increase the performance and safeguard surface pumping systems. The application determines the components of the surface pumping system.

(*Continued*)

Table 12.3 Electric water pumping categories [25, 26]. (*Continued*)

Type	Major components	Role of components
Submersible Pumps	• Electric Motor • Pump Impeller • Pump Housing • Inlet Screen • Cable • Control Panel • Protection Devices • Discharge Connection	Submersible pumps have the following main parts: 1. **Electric Motor**: Provides the power to drive the pump and is sealed to prevent water from entering 2. **Pump Impeller**: Use to move the water by creating a flow and generating pressure 3. **Pump Housing**: Encloses the pump impeller and its operating parts. It can be immersed 4. **Inlet Screen**: Protects the pump's impeller from foreign objects. 5. **Cable:** Use to integrate the electric motor to the control panel or power source. 6. **Control Panel:** The wiring, fuses, and switches for the pump's electrical system are found here. 7. **Protection Devices:** Pressure switches, dry-run prevention, and overload protection are all in place to ensure that the pump runs smoothly. 8. **Discharge Connection:** Connects the pump to the discharge piping.

(*Continued*)

Table 12.3 Electric water pumping categories [25, 26]. (*Continued*)

Type	Major components	Role of components
Centrifugal Pumps	• Impeller • Shaft • Casing • Suction and discharge nozzle • Mechanical seal • Bearings • Motor system • Volute	1. **Impeller:** spins to enhance liquid pressure and flow. 2. **Shaft:** Transfers the power from the motor to the impeller. 3. **Casing:** encloses the impeller and seals the pump to prevent leakage. 4. **Suction and discharge nozzle:** Connect the pump to the inlet and outlet piping 5. **Mechanical seal:** Prevents liquid from escaping between the shaft and casing 6. **Bearings:** Support the shaft and allow it to rotate 7. **Motor:** Provides the power to drive the impeller. 8. **Volute:** Converts the velocity of the fluid into pressure
Positive Displacement Pumps	• Rotor • Stator • Drive mechanism • Suction and discharge connections • Seals • Bearings • Lubrication system • Valve	1. **Rotor:** help for spinning piece that traps and delivers a set amount of fluid to the discharge outlet 2. **Stator:** Fixed device that seals the fluid flow around the rotor 3. **Drive mechanism:** An electric motor powers the rotor. 4. **Suction and discharge connections:** pump liquids in and out 5. **Seals:** prevent fluid leakage between rotor and stator or pump and process. 6. **Bearings:** help the rotor spin smoothly. 7. **Lubrication system:** lubricates bearings and seals to minimize wear. 8. **Valve:** regulates pump fluid flow

(*Continued*)

Table 12.3 Electric water pumping categories [25, 26]. (*Continued*)

Type	Major components	Role of components
Multi-Stage Pumps	• Impellers • Shaft • Casing • Intermediate discharge casings • Suction and discharge nozzle • Mechanical seal • Bearings • Motor • Volute • Diffusers	1. **Impellers:** spinning devices that enhance liquid pressure and flow. 2. **Shaft:** Motor power transferred to impellers 3. **Casing:** plugs the impellers and pump to prevent leaking. 4. **Intermediate discharge casings:** link pump stages to facilitate fluid transmission. 5. **Suction and discharge nozzle:** attach the pump to the inlet/outlet pipework. 6. **Mechanical seal:** Stops liquid flow between shaft and casing 7. **Bearings:** Support the shaft and grant it to move rotationally. 8. **Motor:** supplies the energy necessary to turn the impellers 9. **Volute:** Converts the fluid velocity into pressure 10. **Diffusers:** guide the flow of fluids and transform the energy of their motion into the force of gravity.

12.5 Challenges of Hybrid PV/PEMFC Technology

A possible strategy for long-term energy security is the hybrid PV/PEMFC technology. Despite its benefits, it still faces several obstacles that must be overcome before it can be widely used as,

- **Integration of multiple energy sources:** As each technology has its own set of parameters under which it functions, combining them into a single system may be difficult.
- **Cost reduction**: Because hybrid systems require more complicated components and are generally more expensive to manufacture, they are more expensive than single-source PV or PEMFC systems [27].

- **Efficiency**: Owing to the energy losses incurred during the conversion of one type of energy to another, hybrid systems may be less efficient than single-source systems.
- **Durability**: However, the rigorous working conditions of fuel cells may have an adverse effect on the efficiency and lifespan of solar cells, which might cause concern for hybrid systems.
- **Maintenance complexity**: More parts and systems mean more potential points of failure in a hybrid system than in a single-system setup.

Despite these drawbacks, hybrid PV/PEMFC technology has significant potential for supplying sustainable and dependable energy in a wide range of applications, including distant power systems and residential residences [28].

12.5.1 Challenges of Hydrogen Production and Storage

Maintaining a functional FC system requires large quantities of hydrogen generation and storage. Potential obstacles to hydrogen generation and storage include the following,

1. **Hydrogen production**: Production using a variety of methods, including steam methane reforming, electrolysis, and biological processes. Negative aspects of various methods include, but are not limited to, wasteful use of resources and potentially harmful consequences on the natural world.
2. **Hydrogen storage**: It is flammable and must be stored in specialized containers. In this context, the use of cryogenic storage, metal hydrides, or high-pressure tanks is possible, but each has its own advantages and disadvantages [28].
3. **Distribution**: The absence of a reliable distribution network increases the difficulty of transporting hydrogen from its generation point to its storage point.
4. **Safety**: Extremely flammable improper transport, and storage provide fire hazards.
5. **Producing cost**: Hybrid PV/FC power-generating systems are costlier than other renewable energy systems [29].

Hydrogen is an attractive energy carrier that may play a crucial role in shifting to a low-carbon energy system. Hybrid PV/FC power generating systems may benefit from using hydrogen as a source of energy if hydrogen production and storage technologies were further developed and improved.

12.5.2 Challenges of the Hybrid PV/PEMFC System Integration

However, several obstacles must be overcome to successfully integrate a PV/PEMFC system for hybrid power production.

1. **Compatibility:** It may be difficult to ensure that PV and FC systems can function together efficiently because of their varying operating circumstances and needs.
2. **Efficiency:** While integrating several components, the hybrid system is susceptible to energy-conversion losses, which lowers the overall efficiency of the system.
3. **Control and management:** Because of the inherent complexity of coordinating the functioning of both systems, a well-developed control and management system is required to achieve peak efficiency.
4. **Integration cost:** Combining two technologies into one may drive up the price of that technology owing to the need for more complicated and specialized sub-devices, e.g., relay and control units.
5. **High maintenance liability:** Maintenance of a hybrid system can be more complex than maintaining a standalone system, as there are more components to monitor and maintain from environmental effects, e.g., dust accumulation and non-uniform irradiance due to partial showing conditions.

To overcome these obstacles and produce more efficient hybrid power generation systems, it is necessary to continue developing and improving integration technologies and processes.

12.5.3 Hybrid PV/FC Power System Ignorance and Acceptance

The low rates of adoption of hybrid PV/PEMFC power systems may be caused by a lack of public knowledge about these systems. As a result, human acceptance might be poor, as people are often unwilling to adopt technology that they do not fully comprehend [30]. Educational efforts that highlight the merits and usefulness of hybrid PV/FC systems are crucial for resolving these difficulties. Furthermore, proving dependability and cost-effectiveness in real-world scenarios may boost public adoption.

12.6 Future Scope of Hybrid PV/PEMFC Water-Pumping Systems

The future scope of the hybrid PV/PEMFC-driven water-pumping system for agricultural applications is promising. As technology improves and costs decrease, it has the potential to become more widely adopted in the agricultural sector. The increasing demand for sustainable agriculture and the need for water in agriculture make it an important area of research and development [31].

In the future, improvements in fuel-cell technology and the development of more efficient and cost-effective hydrogen storage solutions are expected to make this system more accessible to farmers. Additionally, advancements in renewable energy technologies and the integration of smart grid systems may lead to more sophisticated and automated water-pumping systems in agriculture.

Overall, the future of the hybrid PV/PEMFC-driven water-pumping system in agriculture appears promising, and this technology has the potential to play a significant role in promoting sustainable agriculture and improving water management practices [32]. Further research and development in this field will help realize its full potential and make this technology more accessible and effective for farmers.

a) **Cost reduction:** With technological breakthroughs and economies of scale, the FC technology costs are likely to continue to fall. The PV system is more acceptable in the current scenarios because it is the most economical source of energy.
b) **Hydrogen infrastructure and sun irradiance availability:** Hydrogen refueling infrastructure will develop because of public and private sector investments in hydrogen-filling stations. Hybrid power production may also benefit from the sun's irradiance, which can be used to generate energy from a photovoltaic system and serve as a backup power source.

c) **Government support:** Government support policies for hybrid PV/PEMFC powered water-pumping systems differ greatly from one nation to the next. The installation of renewable energy systems such as hybrid PV/FC water-pumping systems may be subsidized or given tax breaks in certain nations. Indirect support for the use of hybrid PV/FC water-pumping systems may be rules or standards that encourage the use of renewable energy in other nations. Consult with professionals in the fields of renewable energy and water-pumping systems and research the unique regulations and incentives available in the country.
d) **Demand due to high reliability:** Several variables may affect how long a hybrid PV/PEMFC water-pumping system lasts, including the quality of its components, how well it is maintained, and how well it was designed. Hybrid systems that use both PV and FC technologies are more reliable than the technologies used alone. In times of poor solar insolation, the FC component may kick in and provide backup power for the water pump. When solar insolation levels are sufficiently high, the PV system generates renewable energy and reduces the use of fossil fuels.

12.7 Pros and Cons of Hybrid PV/PEMFC-Powered Water-Pumping System

The pros and cons of the hybrid PV/PEMFC-driven water-pumping system for agricultural applications are important because they highlight the benefits and drawbacks of using this technology for irrigation [33]. The pros of the system include reliability, increased efficiency, sustainability, and improved water management, which can lead to increased crop yields and reduced water waste. However, the cons include the high cost of FCs, need for regular maintenance, difficulties in storing hydrogen fuel, lack of standardization, and complexity of the system [34].

It is essential to weigh these pros and cons when considering the use of this technology in agriculture, as they can impact the overall effectiveness and feasibility of the system [35]. Understanding the trade-offs involved in implementing a hybrid PV/PEMFC-driven water-pumping system can help farmers make informed decisions about whether this technology is appropriate for their needs. Some of the pros and cons of hybrid PV/PEMFC power-assisted water-pumping system included in the Table 12.4 as,

Table 12.4 Pros and cons of PV/PEMFC-powered water-pumping system.

Pros	
Reliable energy source	• The hybrid photovoltaic/PEM fuel cell system delivers dependable water pumping energy, particularly in distant and off-grid situations
Increased efficiency	• The fuel cell can generate electricity when the sun is not shining, increasing system efficiency
Sustainable	• Renewable energy decreases dependency on fossil fuels, making this system more sustainable and ecologically benign
Improved water management	• Renewable energy reduces fossil fuel dependence, making this system more sustainable and eco-friendlier
Cons	
High cost	• FCs are costly, making the hybrid PV/PEMFC system more expensive than alternative water pumping systems
Maintenance	• FC maintenance is costly and complicated
Hydrogen storage	• Hydrogen fuel storage is difficult and needs specific equipment, which increases system cost
Lack of standardization	• Lack of standards and technical understanding makes it hard for farmers to apply these systems
Complexity	• Farmers may struggle to utilize and maintain the hybrid PV/PEMFC system due to its complexity

12.8 Conclusion

A hybrid PV/PEMFC-driven water-pumping system for agricultural applications combines two sources of renewable energy, solar PV and PEMFCs, to power a water pump for irrigation purposes. The proposed system offers the advantage of providing a reliable source of energy for water pumping in remote and off-grid locations. However, there are several challenges associated with this technology, including the high cost of FCs, need for regular maintenance, and difficulty in storing hydrogen fuel. As technology improves and costs decrease, it is likely that it will become more widely adopted. The increasing demand for sustainable agriculture and the need for water in agriculture make this an important area of research and development. In addition, hybrid PV/PEMFC-driven water-pumping systems are promising solutions for sustainable agriculture, but further research and development are needed to overcome the current challenges and make this technology more accessible to farmers. Moreover, the hydrogen storage for PEMFC operation is also explored, along with its benefits, drawbacks, and potential for improvement, in this research

In this study, the transient behavior of a hybrid PV/PEMFC power-driven water-pumping system was examined through mathematical analysis and MATLAB/Simulink modeling. In addition, the water pumping motor is driven at 1,728 rpm at a hybrid PV/FC voltage of 810 V.

References

1. Tintanel, M. T., Gracia A. M., Uche J., Amo, A. D., Rujula A. A., B., Uson S., Arauzo, I. (2022). Energy and environmental benefits of an integrated solar photovoltaic and thermal hybrid, seasonal storage and heat pump system for social housing. *Appl. Therm. Eng.*, 213, 1-13.
2. Malleswararao, K., Aswin, N., Kumar, P., Dutta, P., Murthy, S. S. (2022). Experiments on a novel metal hydride cartridge for hydrogen storage and low temperature thermal storage. *Int. J. Hydrogen Energy*, 47(36), 16144-16155.
3. Sanguesa, J. A., Torres-Sanz, V., Garrido, P., Martinez, F. J., & Marquez-Barja, J. M. (2021). A review on electric vehicles: Technologies and challenges. *Smart Cities*, 4(1), 372-404.
4. Du, J., Ouyang, M., & Chen, J. (2017). Prospects for Chinese electric vehicle technologies in 2016–2020: Ambition and rationality. *Energy*, 120, 584-596.
5. Van Mierlo, J., Berecibar, M., El Baghdadi, M., De Cauwer, C., Messagie, M., Coosemans, T., Hegazy, O. (2021). Beyond the state of the art of electric vehicles: A fact-based paper of the current and prospective electric vehicle technologies. *World Electr. Veh. J.*, 12(1), pp. 1-26.

6. Sharma R., Sharma S., Tiwari S. (2020). Design optimization of sola. PV water pumping system. *Mater. Today: Proc.*, 21(3), 1673-1679.
7. Pachauri, R. Sharma, A. Rajput, S., (2020). Parametric effects on proton exchange membrane fuel cell performance: an analytical perspective. *Int. J. Math. Eng. Manag. Sci.*, 5(5) 926-938.
8. Rupendra Kumar Pachauri & Yogesh K. Chauhan (2017) Study and performances analysis of fuel cell assisted vector control variable speed drive system used for electric vehicles. *Int. J. Sustain. Energy*, 36:2, 167-191, DOI: 10.1080/14786451.2015.1017495
9. Dimitrova, Z., Nader, W. B. (2022). PEM fuel cell as an auxiliary power unit for range extended hybrid electric vehicles. *Energy*, 239, pp. 1-12.
10. Pachauri, R., Chauhan, Y. K., (2015). A Study, Analysis and Power Management Schemes for Fuel Cells," *Renew. Sust. Energ. Rev. (Elsevier)*, 43, 1301-1319.
11. Pachauri, R. K., Chauhan, Y. K., Gupta, A., (2016). Proton Exchange Membrane Fuel Cell: Investigation of Control Schemes for Reactants Flow Pressure. *Energy Sources Part A: Recovery Util. Environ. Eff.*, 39(1), 75-82.
12. Sid, M. N., Becherif, M., Aboubou, A., Benmouna, A. (2022). Power control techniques for fuel cell hybrid electric vehicles: A comparative study. *Comput. Electr. Eng.*, 97, pp. 1-19.
13. Pachauri, R., Chauhan, Y. K., Kazmi, S. A., (2020). Power management system for isolated hybrid PV/WT/PEMFC power generation. *Int. J. Sci. Technol. Res.*, 9(2), 4756-4766.
14. Khan, A.A.; Minai, A.F.; Pachauri, R.K.; Malik, H. Optimal Sizing, Control and Management Strategies for Hybrid Renewable Energy Systems: A Comprehensive Review. *Energies* 2022, 15(17), 6249. https://doi.org/10.3390/en15176249.
15. Kansal, I., Pachauri, R. K., (2021). Mathematical Puzzle-based PV Array configuration for GMP enhancement under non-uniform irradiation. *EAI Endorsed Trans. Energy Web*, 8(32), 1-10.
16. A. F. Minai, M.A. Husain, M. Naseem, A.A. Khan, (2021) "Electricity Demand Modeling Techniques for Hybrid Solar PV System", *Int. J. Emerg. Electr. Power Syst. (IJEEPS)*, ISSN: 1553-779X, Published by De Gruyter July 7, 2021 https://doi.org/10.1515/ijeeps-2021-0085.
17. A. A. Khan, A. F. Minai, L. Devi, Q. Alam and R. K. Pachauri, "Energy Demand Modelling and ANN Based Forecasting using MATLAB/Simulink," *2021 International Conference on Control, Automation, Power and Signal Processing (CAPS)*, 2021, pp. 1-6, doi: 10.1109/CAPS52117.2021.9730746.
18. M. Naseem, Mohammad Aslam Husain, J Dinesh Kumar, Md, Naseem Ahmad, A. F. Minai, Akhlaque Ahmad Khan, "Particle Swarm Optimization based Maximum Power Tracking Techniques for Solar PV System under Partially Shaded Conditions," *2021 International Conference on Control, Automation, Power and Signal Processing (CAPS)*, 2021, pp. 1-6, doi: 10.1109/CAPS52117.2021.9730746.

19. A.A. Khan, A. F. Minai (2023)"A Strategic Review: The Role of Commercially Available Tools for Planning, Modeling, Optimization, and Performance Measurement of Photovoltaic Systems", *Energy Harvesting Syst. (EHS)*, ISSN: 2329-8766. https://doi.org/10.1515/EHS-2022-0157.
20. M. Kumar, A. F. Minai, A. A. Khan and S. Kumar, "IoT based Energy Management System for Smart Grid," *2020 International Conference on Advances in Computing, Communication & Materials (ICACCM)*, 2020, pp. 121-125, doi: 10.1109/ICACCM50413.2020.9213061.
21. A. F. Minai, A. A. Khan, M. A. Siddiqui, R. K. Pachauri and A. Raj, "Design and cost study of a 25kW SPV system based on real performance in an Indian environment" in *fifth International Conference on Intelligent Communication, Control and Devices (ICICCD-2022)* to be held on 11th-12th November, 2022.
22. Malla S. G, Bhende, C. N., Mishra S., (2011). Photovoltaic based water pumping system, in *Proc. IEEE Conference on Energy, Automation and Signal* at Bhubaneswar, India, 1-6.
23. Meah K., Fletcher S., Ula S., (2008). Solar photovoltaic water pumping for remote locations. *Renew. Sust. Energ. Rev.*, 12(2) 472-487.
24. Thakur, A. K., Singh, R., Gehlot, A, Kaviti, A. K., Aseer, R., Suraparaju, S. K., Natarajan S. K., Sikarwar V. S., (2022) Advancements in solar technologies for sustainable development of agricultural sector in India: a comprehensive review on challenges and opportunities. *Environ. Sci. Pollut. Res.* 29, 43607–43634.
25. Belgacem, A., Miloud, Y., Mostefai M., Belgacem, F. (2022) Fuzzy logic direct torque control of induction motor for photovoltaic water pumping system. *Int. J. Power Electron. Drive Syst.*, 13(3), 1822-1832.
26. Hilali A., Mardoude, Y., Essahlaoui A., Rahali A., Ouanjli N. E., (2022) Migration to solar water pump system: Environmental and economic benefits and their optimization using genetic algorithm based MPPT. *Energy Rep.*, 8, 10144-10153.
27. Eren Y., (2022) Uncertainty-aware non-supplied load minimization oriented demand response program for PV/FC power system with electrolyzer back-up. *Int. J. Hydrogen Energy*, 47(19), 10715-10735.
28. Tu, J., Bai, Z., Wu, X. (2022). Sizing of a Plug-In Hybrid Electric Vehicle with the Hybrid Energy Storage System. *World Electr. Veh. J.*, 13(7), 1-10.
29. Rezaei, H., Abdollahi, S. E., Abdollahi, S., Filizadeh, S. (2022). Energy management strategies of battery-ultra-capacitor hybrid storage systems for electric vehicles: Review, challenges, and future trends. *J. Energy Storage*, 53, 1-19.
30. Muralidhar, Krishnappa, and Natarajan Rajasekar. "A review of various components of solar water-pumping system: Configuration, characteristics, and performance." *Int. Trans. Electric Energy Syst.*, 31.9 (2021): e13002. https://doi.org/10.1002/2050-7038.13002

31. Kapen P. T., Nouadje B. A. M., Chegnimonhan V., Tchuen G., Tchinda R., (2022) Techno-economic feasibility of a PV/battery/fuel cell/electrolyzer/ biogas hybrid system for energy and hydrogen production in the far north region of cameroon by using HOMER pro. *Energy Strategy Rev.* 44, 1-25.
32. Babatunde, O. M., Munda J. L., Hamam, Y., (2022) Off-grid hybrid photovoltaic – micro wind turbine renewable energy system with hydrogen and battery storage: Effects of sun tracking technologies. *Energy Convers. Manage.*, 255, 1-17.
33. A. K. Bansal, (2022). Sizing and forecasting techniques in photovoltaic-wind based hybrid renewable energy system: A review. *J. Cleaner Prod.*, 369, 1-23.
34. Hamza B., Abdelghani H., Hegazy R., (2022). Optimal configuration and techno-economic analysis of hybrid photovoltaic/PEM fuel cell power system. *J. New Mater. Electrochem. Syst.* 25(2), 116-125.
35. Ishaq, H., Dincer, I., & Crawford, C. (2022). A review on hydrogen production and utilization: Challenges and opportunities. *Int. J. Hydrogen Energy*, 47(62), 26238-26264.

Index

Also of Interest

Check out these other related titles from Scrivener Publishing

CLEAN AND RENEWABLE ENERGY PRODUCTION, Edited by Surajit Mondal, Adesh Kumar, Rupendra Kumar Pachauri, Amit Kumar Mondal, Vishal Kumar Singh, and Amit Kumar Sharma, ISBN: 9781394177723. This exciting new volume explores the recent advances in renewable energy production, the policies surrounding renewable energy, and practical applications.

URBAN ENERGY SYSTEMS: Modelling and Simulation for Sustainable Smart Cities, Edited by Deepak Kumar, ISBN: 9781119847441. Presenting the latest technologies and practices in this ever-changing field, this groundbreaking new volume covers the gambit for providing solutions and practical applications of smart and efficient energy systems.

ENERGY STORAGE TECHNOLOGIES IN GRID MODERNIZATION, Edited by Sandeep Dhundhara, Yajvender Pal Verma, and Ashwani Kumar, ISBN: 9781119872115. Written and edited by a team of experts, this exciting new volume discusses the various types of energy storage technologies, the applications of energy storage systems, the performance improvement of modern power systems, their role in the real-time operation of power markets, and the operational issues of modern power systems, including renewable-based generating sources.

SMART GRIDS FOR SMART CITIES VOLUME 1, Edited by O.V. Gnana Swathika, K. Karthikeyan, and Sanjeevikumar Padmanaban, ISBN: 9781119872078. Written and edited by a team of experts in the field, this first volume in a two-volume set focuses on an interdisciplinary perspective on the financial, environmental, and other benefits of smart grid technologies and solutions for smart cities.

SMART GRIDS FOR SMART CITIES VOLUME 2: Real-Time Applications in Smart Cities, Edited by O.V. Gnana Swathika, K. Karthikeyan, and Sanjeevikumar Padmanaban, ISBN: 9781394215874. Written and edited by a team of experts in the field, this second volume in a two-volume set focuses on an interdisciplinary perspective on the financial, environmental, and other benefits of smart grid technologies and solutions for smart cities.

SMART GRIDS AND INTERNET OF THINGS, Edited by Sanjeevikumar Padmanaban, Jens Bo Holm-Nielsen, Rajesh Kumar Dhanaraj, Malathy Sathyamoorthy, and Balamurugan Balusamy, ISBN: 9781119812449. Written and edited by a team of international professionals, this groundbreaking new volume covers the latest technologies in automation, tracking, energy distribution and consumption of Internet of Things (IoT) devices with smart grids.

SMART GRIDS AND GREEN ENERGY SYSTEMS, Edited by A. Chitra, V. Indragandhi and W. Razia Sultana, ISBN: 9781119872030. Presenting the concepts and advances of smart grids within the context of "green" energy systems, this volume, written and edited by a global team of experts, goes into the practical applications that can be utilized across multiple disciplines and industries, for both the engineer and the student.

SMART GRIDS AND MICROGRIDS: Concepts and Applications, Edited by P. Prajof, S. Mohan Krishna, J. L. Febin Daya, Umashankar Subramaniam, and P. V. Brijesh, ISBN: 9781119760559. Written and edited by a team of experts in the field, this is the most comprehensive and up to date study of smart grids and microgrids for engineers, scientists, students, and other professionals.

MICROGRID TECHNOLOGIES, Edited by C. Sharmeela, P. Sivaraman, P. Sanjeevikumar, and Jens Bo Holm-Nielsen, ISBN 9781119710790. Covering the concepts and fundamentals of microgrid technologies, this volume, written and edited by a global team of experts, also goes into the practical applications that can be utilized across multiple industries, for both the engineer and the student.

INTEGRATION OF RENEWABLE ENERGY SOURCES WITH SMART GRIDS, Edited by A. Mahaboob Subahani, M. Kathiresh and G. R. Kanagachidambaresan, ISBN: 9781119750420. Provides comprehensive coverage of renewable energy and its integration with smart grid technologies.

Encyclopedia of Renewable Energy, by James G. Speight, ISBN: 9781119363675. Written by a highly respected engineer and prolific author in the energy sector, this is the single most comprehensive, thorough, and up to date reference work on renewable energy.

Green Energy: Solar Energy, Photovoltaics, and Smart Cities, edited by Suman Lata Tripathi and Sanjeevikumar Padmanaban, ISBN: 9781119760764. Covering the concepts and fundamentals of green energy, this volume, written and edited by a global team of experts, also goes into the practical applications that can be utilized across multiple industries, for both the engineer and the student.

Energy Storage, edited by Umakanta Sahoo, ISBN: 9781119555513. Written and edited by a team of well-known and respected experts in the field, this new volume on energy storage presents the state-of-the-art developments and challenges in the field of renewable energy systems for sustainability and scalability for engineers, researchers, academicians, industry.

Energy Storage 2nd Edition, by Ralph Zito and Haleh Ardebili, ISBN: 9781119083597. A revision of the groundbreaking study of methods for storing energy on a massive scale to be used in wind, solar, and other renewable energy systems.

Hybrid Renewable Energy Systems, edited by Umakanta Sahoo, ISBN: 9781119555575. Edited and written by some of the world's top experts in renewable energy, this is the most comprehensive and in-depth volume on hybrid renewable energy systems available, a must-have for any engineer, scientist, or student.

Progress in Solar Energy Technology and Applications, edited by Umakanta Sahoo, ISBN: 9781119555605. This first volume in the new groundbreaking series, Advances in Renewable Energy, covers the latest concepts, trends, techniques, processes, and materials in solar energy, focusing on the state-of-the-art for the field and written by a group of world-renowned experts.

A Polygeneration Process Concept for Hybrid Solar and Biomass Power Plants: Simulation, Modeling, and Optimization, by Umakanta Sahoo, ISBN: 9781119536093. This is the most comprehensive and in-depth study of the theory and practical applications of a new and groundbreaking method for the energy industry to "go green" with renewable and alternative energy sources.

DESIGN AND DEVELOPMENT OF EFFICIENT ENERGY SYSTEMS, edited by Suman Lata Tripathi, Dushyant Kumar Singh, Sanjeevikumar Padmanaban, and P. Raja, ISBN: 9781119761631. Covering the concepts and fundamentals of efficient energy systems, this volume, written and edited by a global team of experts, also goes into the practical applications that can be utilized across multiple industries, for both the engineer and the student.

INTELLIGENT RENEWABLE ENERGY SYSTEMS: Integrating Artificial Intelligence Techniques and Optimization Algorithms, edited by Neeraj Priyadarshi, Akash Kumar Bhoi, Sanjeevikumar Padmanaban, S. Balamurugan, and Jens Bo Holm-Nielsen, ISBN: 9781119786276. This collection of papers on artificial intelligence and other methods for improving renewable energy systems, written by industry experts, is a reflection of the state of the art, a must-have for engineers, maintenance personnel, students, and anyone else wanting to stay abreast with current energy systems concepts and technology.

SMART CHARGING SOLUTIONS FOR HYBRID AND ELECTRIC VEHICLES, edited by Sulabh Sachan, Sanjeevikumar Padmanaban, and Sanchari Deb, ISBN: 9781119768951. Written and edited by a team of experts in the field, this is the most comprehensive and up to date study of smart charging solutions for hybrid and electric vehicles for engineers, scientists, students, and other professionals.

Printed in the USA/Agawam, MA
September 16, 2026

910661.063